Microbes as Biofertilizers and their Production Technology

Microbes as Biofertilizers and their Production Technology

Prof. S. G. Borkar

WOODHEAD PUBLISHING INDIA PVT LTD

New Delhi, India

Contents

Preface   xiii

Foreword   xv

1   Introduction   **1**

2   Microbes as biofertilizers, their types and importance   **7**

    2.1   What are biofertilizers?   7

    2.2   Types of biofertilizers   7

        2.2.1   Nitrogen-fixing biofertilizers   8

        2.2.2   Phosphate solubilizing biofertilizers   39

        2.2.3   Vesicular arbascular mycorrhizae   43

        2.2.4   Potassium solubilizing biofertilizer   46

        2.2.5   Sulfur oxidizing biofertilizer   50

        2.2.6   Silicate solubilizing biofertilizer   54

        2.2.7   Silica solubilizing bacillus sp.   58

        2.2.8   Decomposting cultures   59

3   Rhizobium   **63**

    3.1   Introduction   63

    3.2   Isolation of *Rhizobium*   63

        3.2.1   Materials   64

        3.2.2   Procedure   65

        3.2.3   Observations   66

    3.3   Cultural tests to distinguish Rhizobia
from contaminants   67

    3.4   Estimation of 'N' fixing efficiency of
Rhizobium cultures   68

        3.4.1   Apparatus   69

        3.4.2   Procedure   70

        3.4.3   Calculation   70

4   Azotobacter   **73**

4.1   Introduction   73

4.2   Isolation of *Azotobacter* from soil on selective
        Jonsen's agar medium   73

        4.2.1   By soil dilution and pour plate technique   73

        4.2.2   By enrichment culture technique   75

4.3   Cultural test to distinguish *Azotobacter* from
        contaminant   76

4.4   Test for Nitrogen fixation in pure culture
        of *Azotobacter*   77

        4.4.1   Pure culture medium   77

5   Azospirillum   **79**

5.1   Introduction   79

5.2   Isolation of *Azospirillum* from soil   79

        5.2.1   Procedure   79

5.3   Cultural test to distinguish *Azospirillium* from
        contaminant   80

5.4   Test for Nitrogen fixation in pure culture
        of *Azospirillum*   80

5.4.1   Pure culture medium   81

6   *Acetobacter diazotrophicus*   **83**

6.1   Isolation of *Acetobacter diazotrophicus*   83

        6.1.1   Materials   83

        6.1.2   Procedure   84

        6.1.3   Observations   84

6.2   Cultural tests to distinguish *Acetobacter
        diazotrophicus* from contaminatns   85

6.3   Test for nitrogen fixation in pure culture of
        Acetobacter diazotrophicus   86

7   Blue-green algae   **87**

7.1   Isolation of blue-green algae from soil   87

7.2   Cultural tests to distinguish blue-green algae from
        contaminants   88

| 7.3 | Test for nitrogen fixation in pure culture of blue-green algae | 89 |

8   Azolla   **91**

8.1	Introduction	91
8.2	Isolation of *Azolla*	91
8.3	Factors influencing the biomass production of *Azolla*	93
	8.3.1 Water	93
	8.3.2 Temperature	94
	8.3.3 Humidity	94
	8.3.4 Light intensity	94
	8.3.5 Wind	95
	8.3.6 Soil pH	95

9   *Frankia*: N_2-fixing endophytes   **97**

9.1	Isolation of *Frankia* from actinorhizal nodules	97
9.2	Cultural test to differentiate *Frankia* from contaminats	98
9.3	Estimation of 'N' fixing efficiency of *Frankia*	98

10   Phosphate solubilizing microorganism   **99**

10.1	Isolation of phosphate solubilizing microorganism	99
	10.1.1 Materials	99
	10.1.2 Pikovskaya's medium	99
	10.1.3 Procedure	100
	10.1.4 Observation	100
10.2	Estimation of phosphate solubilizing ability of phospho solubilizer	100
10.3	Quantification of soluble phosphorus released by phosphosolubilizer	101
	10.3.1 By using ammonium molybdate reagent	101
	10.3.2 By using ascorbic acid reagents	101

11   Vesicular arbuscular mycorrhizae   **103**

| 11.1 | Introduction | 103 |

11.2	Isolation of arbuscular mycorrhizal fungal spores	103
11.3	Identification of VAM fungi	104
	11.3.1 Spore morphology of VAM fungi	104
11.4	Inoculum production	105
11.5	Mycorrhizal dependency	106
11.6	Screening for effective isolates	107
11.7	Application of inoculum	107
11.8	Arbuscular mycorrhizal fungal colonization	107
11.9	Estimation of infectious propagule per gram of soil	108
12	**Sulphur oxidizing microorganism**	**109**
12.1	Isolations of sulphur oxidizing bacteria	109
12.2	Purification of sulphur oxidizing bacteria from enriched medium	109
12.3	Isolations, purification and maintenance of sulphur oxidizing fungi and actinomycetes	110
12.4	Screening of isolates for sulphur oxidizing efficacy	110
13	**Silicate solubilizing biofertilizers**	**113**
13.1	Introduction	113
13.2	Isolation and enumeration	113
	13.2.1 Modified bunt and rovira medium	113
13.3	Testing of silicate solubilization potential	114
	13.3.1 Basal medium	114
13.4	Estimation of available silica (Saxena 1989)	114
	13.4.1 Reagents	114
	13.4.2 Procedure	115
14	**Potassium solubilizing biofertilizers**	**117**
14.1	Isolation of potassium solubilizing microorganisms	117
	14.1.1 Media composition	117
14.2	Testing of effective strains of KSM	117
14.3	Quantitative estimation of k released from insoluble k	118
	14.3.1 Preparation of standard curve	118

15   Composting culture   119

15.1   Isolation and maintenance of decomposting cultures   119

    15.1.1   Materials   119

    15.1.2   Procedure   119

    15.1.3   Observations   120

16   Production technology of biofertilizers   123

16.1   Design for biofertilizer production unit   123

    16.1.1   Laboratory   123

    16.1.2   Fermentor room   124

    16.1.3   Mixing and curing room   124

    16.1.4   Packaging room   124

    16.1.5   Storage room   125

16.2   Production technology   125

    16.2.1   Materials   125

    16.2.2   Procedure   125

16.3   Carrier-based biofertilizers   126

    16.3.1   Preparation of carrier   126

    16.3.2   Preparation of inoculants in powder form   126

16.4   Liquid biofertilizer   128

    16.4.1   Production technology of liquid biofertilizers   128

    16.4.2   Composition of liquid biofertilizer media   128

    16.4.3   Procedure for liquid biofertilizer production   130

16.5   Production technology for blue-green algae132

    16.5.1   Mass production of BGA   132

    16.5.2   Recommendation for field application   134

    16.5.3   Importance of Algal biofertilizers   135

16.6   Production technology of *Azolla*   136

16.7   On-farm production of arbuscular mycorrhizal fungi   136

16.8   Production technology of decomposting culture   137

    16.8.1   Materials   137

    16.8.2   Procedure   137

16.9   Production of compost   138

         16.9.1 Materials   138

         16.9.2 Procedure   138

         16.9.3 Precautions   139

17   **Assessment of quality standards of biofertilizers**   **141**

     17.1   Introduction   141

     17.2   Quality standards of *Rhizobium* and Azotobacter   141

     17.3   Quality control measures (as per ISI specification)   143

18   **Application technology of biofertilizers**   **145**

     18.1   Use of carrier-based biofertilzers   145

             18.1.1 Seed treatment   145

             18.1.2 Seedling treatment   145

             18.1.3 Soil treatment   146

     18.2   Use of liquid biofertilizer   146

             18.2.1 Seed Treatment   146

             18.2.2 Sett inoculation   146

             18.2.3 Seedling treatment   147

             18.2.4 Soil treatment   147

             18.2.5 Soil broadcasting   147

             18.2.6 Seed pelleting   147

             18.2.7 Foliar spray   147

             18.28   Drip irrigation   147

19   **Advantages, limitations and constraints**   **149**

     19.1   Advantages of biofertilizers   149

     19.2   Limitations   150

     19.3   Constraints in biofertilizer production technology   150

             19.3.1 Physical and environmental constraints   150

             19.3.2 Chemical constraints   151

             19.3.3 Biological constraints   151

             19.3.4 Technical constraints   151

             19.3.5 Infrastructure constraints   152

19.3.6   Human constraints   152

Glossary   **155**
List of Medium   **159**
References   **165**
Subject index   **195**
Author index   **197**

Preface

Agriculture plays a crucial role in Indian economy, accounting close to 17% of gross domestic product. More importantly, about 60% of India's work force is dependent on agriculture and allied activities for their livelihood. Successive Five Year Plans on self-sufficiency and self-reliance in food grain production have resulted in substantial increase in agricultural production and productivity. The use of chemical fertilizers has played a vital role in the success of India's green revolution and consequent self-reliance in food grain production.

To achieve this, the fertilizer consumption in India increased from less than 1 kg/hector in 1951–1952 to 143.130 kg/hector in 2011–2012. The indigenous capacity of fertilizer production which was around 27 lakh metric tonnes during 2011–2012 in nutrient terms, remained almost stagnant for over more than one decade; and therefore, the government had to increase the import during 2013–2014 which was estimated to be around 80 lakh metric tonnes (80 LMT). Around 25% of our fertilizer requirements are met through this import. This increasing dependence on imports had serious implication on the economy.

This dependence on inorganic fertilizers can be reduced by using the organic fertilizers like farmyard manure, compost, green manure, slurries, and more importantly biofertilizers. Biofertilizers can be used almost in all the crop plants, i.e. hydrophytes like paddy and sugarcane, mesophytes like most of the cereal, pulses, vegetables, and fruit crops, and zerophytes like pearl millet and medicinal plants. The research findings available show that the biofertilizers like *Acetobacter* can save up to 50% of nitrogen application in sugarcane, while in several other crops it can saves 20% nitrogenous application and increase the yield by 20%. When we considered the estimated import of nitrogenous fertilizers urea of 80 LMT, at least by application of biofertilizers, the import of 16 LMT of nitrogenous fertilizer can be reduced. Similarly with the use of P solubilzer, K mineralizer, and S oxidizer, the requirement of these nutrients can be accomplished. With the use of organic sources like compost and biofertilizers, not only the nutrient supplements can be made available in soil but also the structure, texture, and soil health can be improved which has long implication on soil health and productivity.

In the present day scenario, with continuous use of inorganic fertilizers, the soil fertility is eroded with changed physical/chemical structures and textures, accumulation of toxicants in soil, degraded soil biota and soil health with stagnant soil productivity. Therefore, the need of the day is to have integrated nutrient source system which includes inorganic, organic, and biofertilizers in all the crop production systems. As biofertilizers are cost effective, ready to use, and sustainable in soil ecological systems, their use should be encouraged in all the crop production systems to produce more healthy and organic food.

Prof. S. G. Borkar

Foreword

The year 2015 has been declared by the United Nations as the International Year of the Soil. This is in recognition of the important role soil fertility and productivity play in meeting the Zero Hunger Challenge of the United Nations. When I was the chairman of the FAOs Food Security Committee, I got a Global Soil Partnership organised on the lines of the already existing Global Water Partnership. In our country, land is a shrinking resource for agriculture. Therefore, we have no option except to produce more from diminishing per capita availability of arable land.

Soil hunger leads to crop hunger and thereby to human hunger. To provide all the nutrients needed for our crops through mineral fertilisers, we will require nearly 30 million metric tonnes of fertilisers. Such a large quantity will both be difficult to obtain and also ecologically not desirable. Therefore, we should promote integrated nutrient supply systems consisting of a mixture of organic manures like compost and cow dung manures, green manures, biofertilizers, and the minimum essential mineral fertilizers. It is in this context that the present book by Prof. S. G. Borkar and Dr. M. H. Shete of the Department of Agricultural Microbiology of Mahatma Phule Agriculture University is a timely contribution. Efficient microorganisms (also known as EM technology) are well known to be the main instruments for achieving an evergreen revolution leading to increase in productivity in perpetuity without ecological harm.

I hope this book will be widely read and used. Biofertilizers and agricultural microbiology should occupy an important place in the programmes for the international year of soil. We thank Dr. Borkar and Dr. Shete for this labour of love in the cause of sustainable food security in our country.

M. S. Swaminathan

Introduction

A large proportion of humanity depends for its sustenance on the food production to which increase is brought about through the application of fertilizers to the crops. Fertilizers contribute to both the quantity and quality of the food produce when used in the right way; applying the right source at right rate, time and place and to the right crops (http://www.fertilizer.org).

Fertilizers are as important as the seed; as evident in the green revolution (Tomich et. al. 1995), contributing as much as 50% of the yield growth in Asia (Hopper 1993 and FAO 1998). Others have found that one third of cereal production worldwide was due to the use of fertilizers and related factors of production (Bumb 1995). The global fertilizer industry produces some 170 million tonnes of fertilizer nutrients annually. These are used in every corner of the globe to support agricultural production and it is an essential ingredient in the drive towards food security (IFA 2014).

Plants require three major and thirteen minor nutrients for their growth. Nitrogen, phosphorus and potassium are the three major plant nutrients and are supplied through inorganic fertilizers to the plants. It is estimated that annually 25.1 million tonnes of nutrients are removed from the soil, whereas only 15.0 million tonnes are supplied to soil including organics. The total demand of fertilizers as plant nutrient by the whole country cannot be made due to their unavailability. Import of fertilizers in India was about 2 million tonnes in early parts of the year 2000, which increased to 10.2 million tonnes in 2008–2009. India has the second largest fertilizer consumption in the world after China, consuming about 26.5 million tonnes every year. However, the average intensity of fertilizer use in India remains much lower than most countries in the world (Jaga and Patel 2012). The fertilizer consumption in India has increased from less than 1 kg per hector in 1951–1952 to 143.130 kg in 2011–2012. The indigenous capacity of fertilizer production which is presently 27 lakh metric tonnes in nutrient terms remained almost stagnant for more than one decade. The fertilizer production in our country does not fulfill the required demand (Table 1).

Table 1 Status of fertilizer requirements and production in India
(million tonnes)

Year	1996	2000	2011	2031	2051
Requirement	16.0	19.0	20.2	27.3	31.3
Production	12.8	14.9	15.8	20.9	23.9

And therefore, the government is increasing the import every year to meet the domestic requirement. Globally, the consumption of fertilizer-N increased from 8 to 17 kg per hector of agricultural land in the 15 years period from 1973 to 1988 (FAO 1990). Significant growth in fertilizer-N usage has occurred in both developed and developing countries (Peoples et. al. 1995). Import of urea as fertilizer-N in 2013–2014 is estimated to be around 80 lakh metric tonnes, roughly 25% of our requirement of urea alone. Around 90% of P and 100% of K requirements are met through import. Increasing dependence on import has serious implication on our economy. The requirement for fertilizer-N are predicated to increase further in the future (Subba-Rao 1980); however, with the current technology for fertilizer production and the inefficient methods employed for fertilizer application, both the economic and ecological cost of fertilizer usage will eventually become prohibitive. The problems associated with the use of chemical fertilizers are the adverse effects on the long term soil fertility, soil productivity and environmental safety (Kannaiyan 2000). Under these circumstances, alternate sources of fertilizers are to be explored.

The atmosphere contains 78% of free nitrogen which accounts for about 10^{15} tonnes of N_2 gas and the nitrogen cycle involves the transformation of some 3×10^9 tonnes of N_2 per year on a global basis (Postgate 1982). However, transformations (N_2 fixation) are not exclusively biological. World production of nitrogen from dinitrogen for chemical fertilizers accounts for about 25% of the earth's newly fixed N_2, and biological processes account for about 60%. For more than 100 years, biological nitrogen fixation (BNF) has commanded the attention of the scientists concerned with plant mineral nutrition, and it has been exploited extensively in agricultural practice (Burris 1994; Dixon and wheeler 1986). However, its importance as a primary source of N for agriculture has diminished in the recent decades as increasing amounts of fertilizer N have been used for production of food and cash crop (Peoples et. al. 1995). However, international emphasis on environmentally sustainable development with the use of renewable resources is likely to focus attention on the potential role of BNF in supplying N for agriculture (Burris 1994; Peoples et. al. 1995).

Currently, the subject of BNF is of great practical importance because the use of nitrogenous fertilizers has resulted in unacceptable levels of water pollution (increasing concentrations of toxic nitrates in drinking water supplies) and the eutrophication of lakes and rivers (Al-Sherif 1998; Sprent and Sprent 1990). Further, it is usually applied in a few large doses and up to 50% of which may be leached (Sprent and Sprent 1990). This not only wastes energy and money but also leads to serious pollution problems, particularly in water supplies.

To fill this gap, alternate sources of nutrients have to be looked for. Organic wastes and biofertilizers are the alternate sources to meet the nutrient requirement of crops and to bridge the future gaps. Further knowing the deleterious effects of using only chemical fertilizers on soil health, the use of chemical fertilizers along with organic wastes and biofertilizers will be an environment friendly approach for nutrient management and ecosystem maintenance. Such integrated approach will help to maintain the soil health and productivity, and thus should be advocated and popularized among the farming community.

The availability and utilization of the crop nutrients is more important for the crop plants. The crops assimilate their requirement mainly through soil reserve, fertilizers and manure, precipitation and irrigation water and biological nitrogen fixation involving a number of divers groups of microorganisms. Biological nitrogen fixation offers an economically attractive and ecologically sound route for augmenting nutrient supply with the aid of microorganisms. Thus, *Rhizobium* for legumes; BGA and *Azolla* for several crops; P solubilizers, sulphur oxidizers, phosphate absorbers like vesicular arbuscular mycorrhiza (VAM) and organic decomposers plays a significant role in agriculture. All these are based on renewable energy sources and are eco-friendly. But one needs to be clear about the idea that biofertilizers alone can't meet the total nutrient requirement of agriculture and in fact, it is one of the inputs that is to be used along with other inputs.

Biofertilizer is a large population of a specific or a group of beneficial micro-organisms; millions and billions of them are incorporated aseptically into sterile carrier materials such as peat, lignite, or charcoal. Such material is thereafter packed and sold to the farmers as biofertilizers for enhancing productivity of soil, either by fixing atmospheric nitrogen or by solubilising soil phosphorous or by stimulating plant growth through synthesis of growth promoting substances.

Biofertilisers are the low-cost source of plant nutrients, eco-friendly and have supplementary role with chemical fertilizers. The biofertilizers are bacteria, algae and fungi and may broadly be classified into two categories, viz. nitrogen fixing like *Rhizobium*, *Azotobactor*, *Azospirilum*, *Acetobacter*, blue-green algae and *Azolla* and phosphorous solubilizers/mobilisers like PSM and mycorrizae. Recently, the potash mobilisers like *Frateuria aurentia* zinc and sulphur solubilisers like *Thiobacillus* species and manganese solubiliser fungal culture like *Pencillium citrinum* have also been identified for commercial operations. These new strains would also address the issue of "Fertilizer Use Efficiency" and would also enhance the efficacy of biofertilizers.

The biofertilizers were initially identified by a Dutch scientist in 1888 and thereafter "Nobbe and Hiltner" produced the first biofertilizer under the trade name "Nitragin" in 1895 in The United States. The role of biofertilizers in agriculture production assumes special significance, particularly in the present context of very high cost of chemical fertilizers. In simple term, biofertilizers are the fertilizers prepared from bioagents containing live cells of bacteria or cyanobacteria or fungi which efficiently either fix atmospheric nitrogen, solubilize insoluble phosphate in soil, or decompose organic wastes. "Biological nitrogen fixation" is a reduction of molecular nitrogen to ammonia mediated by enzyme nitrogenase. It offers an economically attractive and ecologically sound route for augmenting nutrient supply with the aid of microorganisms.

The biofertilizer production, in general, involves the selection of efficient strain of biofertilizer agent, growing it on a large scale in suitable nutrient medium, incorporating it in sterilized carrier and packing it in biofertilizer packets under strict aseptic conditions.

There are over 100 biofertilizer units operational in the country. These units produced about 20,040 MT biofertilizers against the installed capacity of over 86,000 MT during 2009–2010. The year-wise capacity, production and sales trend of biofertilizers are given in the Table 2.

Table 2   All India production and sales trend of biofertilizers (MT)

Year	Production	Sales
2000–2001	6242.6	6138.6
2001–2002	7390.0	6876.1
2002–2003	8643.0	7000.0
2003–2004	10500.0	8500.0
2004–2005	10479.0	10428.0
2005–2006	11752.0	11358.0
2006–2007	15871.0	15745.0
2007–2008	20111.0	20100.0
2008–2009	25065.0	25000.0
2009–2010	20040.3	20000.0

Source: Specialty Fertilizer Statistics, FAI (2011), MOA and NCOF, Ghaziabad.

The present day requirement of biofertilizers based on crop area for India is 5,50,000 metric tonnes; however, the total production in our country is hardly 20,000 metric tonnes. This necessitates the increase in biofertilizer production. The Government of India is encouraging the biofertilizer industry by providing subsidies to a maximum of 20 lakh rupees and awarding a national productivity award to the efficient biofertilizer production unit. Many state governments have also started self-employment oriented biofertilizer production training programmes. In future, it is hoped that the sufficient production of biofertilizers may be achieved through private, public sectors, and government institutes to meet the requirements.

In contrast to the inorganic fertilizers, biofertlllizers are applied to the seed and seedling and hence they are more effective. They secrete auxins and hormones thereby increasing the germination. Once the biofertilizers are applied, they cannot be leached out, evaporate, or be lost; but the biofertilizer microbes increase in number. Biofertilizer agents also exhibit fungicidal action. Application of biofertilizers does not affect the texture of soil, but improves the structure and texture. Biofertilizers increase nitrogen fixation, phosphate solubilization, sulphur oxidation and helps to supply other macro and micro nutrients to the plants for their growth and are essential for balanced nutrient supply. Similarly they increase vegetative growth, total dry-matter production, yield, and quality of the product. Biofertilizers are the essential soil inputs used in farming that help to develop and sustain soil texture and soil fertility. They are easy to use and can be used for all types of fruits, cereals, pulses, cash crops, vegetables, lawns, ornamental plants, and home-garden plants.

The farmers all over the world are turning more and more towards biological agriculture, as they increasingly encounter the hazardous effects of chemical fertilizers and their prohibitive costs. Biological farming aims at preserving the natural and ecological balance according to the environmental standards.

Therefore, biofertilizer is a most important component in agriculture farming system.

Microbes as biofertilizers, their types and importance

2.1 What are biofertilizers?

The biofertilizers are microbial preparations having manurial value and are applied to seed, soil and seedlings for the enhancement of plant vigor and growth which result in to a higher yield. These are the microbial preparations containing live cells of bacterial, blue-green algae, or fungi, which efficiently fix atmospheric nitrogen, solubilize insoluble phosphate in soil, or decompose cellulosic organic wastes. The efficient strain of the requisite microbial agent is grown on specific nutrient medium on a large scale, aseptically mixed in a suitable carrier and the preparation so formulated is sold as biofertilizer.

Biofertilizers are also called bioinoculants, microbial cultures, bacterial inoculants, or bacterial fertilizers.

2.2 Types of biofertilizers

Biofertilizers are grouped into various categories depending on the kind of beneficial activity carried out by the microbial agent used in their production as follows:

(1) Nitrogen fixing biofertilizers

(2) Phosphate solubilizing biofertilizers

(3) Potassium solubilizing biofertilizers

(4) Sulphur oxidizing biofertilizers

(5) Silicate solubilizing biofertilizers

(6) Decomposing cultures

Carrier-based inoculums of nitrogen fixing, phosphate solubilizing, and decomposing microbial preparations have been commercialized on large scale, but commercial preparations of sulphur oxidizing, silicate solubilizing, potassium solubilizing and VAM biofertilizers have not been yet developed and popularized. Recently, the manufacture of liquid biofertilizers is being geared up and used in the agricultural production system.

2.2.1 Nitrogen-fixing biofertilizers

These include the preparations of bacteria/cyanobacteria which convert atmospheric gaseous nitrogen (N_2) to ammonia and make it available to the plants (Tilak 1993). The organisms which fix atmospheric nitrogen are also called diazotrophs. The nitrogen-fixing organisms include species of *Rhizobium, Azotobacter, Azospirillium, Acetobacter, Azolla* and blue-green algae (cyanobacteria). These are used in commercial production of biofertilizers. The list of nitrogen fixers may be extended to include those diazotrophs which are not being used in commercial production. These are *Beijerinckia indica, Derxia gummose, Clostridium pasteurianum, Klebsiella pneumoniae, Frankia alni, Rhodospirillum rubrum, Rhodospirillum capsulatus, Chlorobium limicola, Chromatium minus, Desulphovibrio* sp and *Methanobacterium* sp. On the basis of association with host, nitrogen-fixing biofertilzers are grouped into following three categories:

(1) Symbiotic N_2 fixing biofertilizers
  e.g., *Rhizobium, Azolla*

(2) Non-symbiotic N_2 fixing biofertilizers
  e.g., *Azotobacter, Beijerinekia, BGA, Acetobacter*

(3) Associative nitrogen fixing biofertilizers
  e.g., *Azospirillum, Frankia*

(1)   Symbiotic N_2 fixing biofertilizers

There is mutual association symbiosis between N_2 fixer and host plants in which they fix atmospheric nitrogen; e.g., Legume-*Rhizobium* symbiosis and *Azolla-Anabaena* symbiosis. Rhizobia produce root nodules in the legumes where nitrogen is fixed symbiotically. The potential role of legumes in maintaining soil fertility is well-known. Grain legumes like green gram, black gram, pigeon pea, lyranthus, etc.; forage crop legumes like clover and alfalfa; vegetable legumes like peas and beans; oilseed legumes like peanuts, soybean, and ornamental plant species of latyrus, lotus and vicia are some of the examples of rhizobial nodulation and symbiotic nitrogen fixation. There are various cross inoculation groups. Each group has specific plant hosts which are colonized by particular species of *Rhizobium* (Table 1).

Table 1 Host specificity of *Rhizobium* strains and per hectare N-fixation

Group	Species	Crops	N_2 fixation kg/ha.
Pea group	*Rhizobium leguminosarum*	Pea, lentil	62–132
Bean group	*Rhizobium phaseoli*	Phaseoli	80–110

Clover group	*Rhizobium trifolii*	Trifolium	130
Alfalfa group	*Rhizobium melioti*	Melilotus	100–150
Soybean group	*Rhizobium japonicum*	Soybean	57–105
Lupini group	*Rhizobium lupine*	Lupinus	70–90
Cicer group	*Rhizobium* sp.	Bengal gram	75–117
Cowpea group	*Rhizobium* sp.	Mung, red gram, Cowpea, groundnut	57–105

Rhizobium

The word rhizobia comes from the ancient Greek word Ρίςα rhiza meaning "root" and Βιος bios meaning "Life," which means life in root. The word *rhizobium* is still sometimes used as a singular form of rhizobia. *Rhizobium* is a genus of gram-negative soil bacteria that fixes atmospheric nitrogen into the plant through root nodules.

The bacteria of genus *Rhizobium* are unicellular, cell size less than 2 μm wide, short to medium rod. It is motile with peritrichous flagella, aerobic, non-spore forming, capsulated. The G + C content of DNA ranges from 59 to 65 moles %. Accumulate poly B-hydroxy butyrate granules. The bacteria are usually present in nodules as pleomorphic forms. Optimum temperature between 25° and 30°C and pH 5.0–8.5 is required for growth; produces leghaemoglobin in root nodules which acts as precursor for N_2 fixation. More the leghaemoglobin in root nodules more is the nitrogen fixing ability of the *rhizobium* strain. Most satisfactory growth is observed on yeast extract mannitol agar media.

Beijerinck in the Netherland was the first to isolate and cultivate a micro-organism from the nodule in 1888. He named it Bacillus radicicola, which is now placed in Bergey's manual of determinative bacteriology under the genus *Rhizobium*.

The first species of rhizobia, *Rhizobium leguminosarum*, was identified in 1889 and all further species were initially placed in the *Rhizobium* genus. Most research has been done on crop and forage legumes such as clover, alfalfa, beans, and soy; recently more work is being done on North American legume.

Rhizobium forms a symbiotic relationship with certain legumes plants and fixes nitrogen from the air into ammonia, which act as a natural fertilizer for the plant. There are 67 species of *Rhizobium* and are specific to crop groups.

Rhizobia are unique in the way that they are the only nitrogen-fixing bacteria living in a symbiotic relationship with legumes. Common crop and forage legumes are peas, beans, clover and soy.

The symbiotic relationship implies a signal exchange between both partners that leads to mutual recognition and development of symbiotic structures. Rhizobia live in the soil where they are able to sense flavonoids secreted by the roots of their host legume plant. Flavonoids trigger the secretion of nod factors which in turn are recognized by the host plant and can lead to root-hair deformation and several cellular responses, such as ion fluxes. The best-known infection mechanism is called intracellular infection. In this case the rhizobia enter through a deformed root-hair in a similar way to endocytosis, forming an intracellular tube called the infection thread. A second mechanism is called "crack entry". In this case, no root-hair deformation is observed and the bacteria penetrate between cells through cracks produced by lateral root emergence. Later on, the bacteria become intracellular and an infection thread is formed like in intracellular infections. The infection triggers cell division in the cortex of the root where a new organ, the nodule, appears as a result of successive processes.

Infection threads grow to the nodule, infect its central tissue and release the rhizobia in these cells where they differentiate morphologically into bacteroids and fix nitrogen from the atmospheric, elemental N_2 into a plant-usable form, ammonium ($NH_3 + H^+ \rightarrow NH_4^+$), using the enzyme nitrogenase. The reaction for all nitrogen-fixing bacteria is:

$$N_2 + 8H^+ + 8e^- \rightarrow 2NH_3 + H_2$$

(http://www.biology.ed.ac.uk)

In return, the plant supplies the bacteria with carbohydrates, proteins and sufficient oxygen so as not to interfere with the fixation process. Leghaemoglobins (plant proteins similar to human hemoglobins) help to provide oxygen for respiration while keeping the free oxygen concentration low enough not to inhibit nitrogenase activity. Recently, a *Bradyrhizobium* strain was discovered to form nodules in Aeschynomene without producing nod factors, suggesting the existence of alternative communication signals other than nod factors (Giraud et. al. 2007).

The legume-*rhizobium* symbiosis is a classic example of mutualism – rhizobia supply ammonia or amino acids to the plant and in return receive organic acids (principally as the dicarboxylic acids malate and succinate) as a carbon and energy source – but its evolutionary persistence is actually somewhat surprising. Because several unrelated strains infect each individual plant, any one strain could redirect resources from nitrogen fixation to its own reproduction without killing the host plant upon which they all depend. But this form of cheating should be equally tempting for all strains, a classic tragedy of the commons. There are two competing hypotheses for the mechanism that maintains legume-*rhizobium* symbiosis (though both may occur in nature). The sanctions hypothesis suggests the plants police cheating rhizobia. Sanctions could take the form of reduced nodule growth, early nodule death, decreased carbon supply to nodules, or reduced oxygen supply to nodules that fix less nitrogen (Denison 2000). The partner choice hypothesis proposes that the plant uses prenodulation signals from the rhizobia to decide whether to allow nodulation and chooses only non-cheating rhizobia. There is evidence for sanctions in soybean plants which reduce *rhizobium* reproduction (perhaps by limiting oxygen supply) in nodules that fix less nitrogen (Kiers et.al. 2003). Likewise, wild lupine plants allocate fewer resources to nodules containing less-beneficial rhizobia, limiting rhizobial reproduction inside. This is consistent with the definition of sanctions just given, although called "partner choice" by the authors (Simms et al. 2006). However, other studies have found no evidence of plant sanctions and instead support the partner choice hypothesis (Heath and Tiffin 2009; Marco et. al. 2009).

The characters which are highly desirable for rhizobial strains to be used in commercial inoculants (Keyser and Li 1992) are as follows:

(1) Ability to form nodules and fix N in target legumes.

(2) Ability to complete nodule formation.

(3) Ability to fix N across a range of environmental conditions.

(4) Ability to from nodules and fix N in the presence of soil nitrate.

(5) Ability to grow well in artificial media, in inoculant carrier and in soil.

(6) Ability to persist in soil, particularly for annually regenerating legumes.

(7) Ability to migrate from initial site of inoculation.

(8) Ability to colonize the soil in the absence of legume host.

(9) Ability to tolerate environmental stresses.

(10) Genetic stability.

(11) Wide host range.

(12) Low mortality on inoculated seed.

(13) Ability to colonize the rhizosphere of host plant.

Habitat: Bacteria are mostly present in the root nodules of legume crops.

Azolla

Azolla (also known as mosquito fern, duckweed fern, fairy moss, or water fern) is a genus of seven species of aquatic fern in the family *Azollaceae*. They are extremely reduced in form and specialized, looking nothing like conventional fern but more resembling duckweed or some mosses. The seven known species are *Azolla circinata*, *A. filiculoides*, *A. Japonica*, *A. Mexicana*, *A. microphylla*, *A. nilotica and A. pinnata while*, *A. rubra* and *A. caroliniana* are the synon of *A. filiculodes* (Hussner 2006). At least six extinct species are known from the fossil record. These are *Azolla intertrappea*, *A. berryi*, *Azolla prisca*, *A. tertiaria*, *A. primaeva and A. boliviensis* (Arnold 1955; Vajda and McLoughlin 2005).

Azolla is a floating heterosporous pteridophyte which contains an endosymbiont Anabaena azollae, a nitrogen fixing cyanobacterium (Nostocaceae family).

Its use as biofertilizer is principally related to the experiments carried out in the Asian countries (Taun and thuyet 1979; Watanabe 1982; Lumpkin and Plucknett 1982). However, the wide range of temperature, humidity and solar radiation in which the fern flourishes suggest the possibility of using *Azolla* outside these regions (Wagner 1997). Carrapico et. al. (2000) also described the use of *Azolla* as biofertilizer in Africa. Bocchi and Malgioglio (2010) reported the use of *Azolla*-anabaena as a biofertilizer for rice paddy field in a temperate rice area in Northern Italy. Three out of five strains which were tested survived the winter with an increase in biomass from March to May, producing approximately 30–40 kg nitrogen per hectare. One of the strains 'Milan' was most resistant to herbicide 'propanil' and was most productive.

The aquatic fern of the genus *Azolla* is small-leaf floating plant which contains an endosymbiotic community living in the dorsal lobe cavity of the pteridophyte leaf. This community is composed of two types of prokaryotic organisms: one species of a nitrogen-fixing filamentous cyanobacteria known as *Anabaena azollae* Strasb. (Described by Strasburger in 1873) and variety of bacteria that few identified as Arthrobacter sp. – an associate with others showing the presence of nitrogenase (Costa et al. 1994). In this association, it is assumed that an exchange of metabolites, namely fixed nitrogen compounds, occurs from the cyanobiont to the host (Carrapico and Tavares 1989a and b).

One of the most interesting features is the role played by the cyanobacterium in this association. Filaments *of Anabaena azollae* are localized in a cavity of the dorsal lobe of the fern's leaves where special conditions stimulate high heterocyst frequency and a vegetative cell differentiation during leaf development (Carrapico and Tavares 1989a). The existence of the two symbionts inside the *Azolla* leaf cavity and its relationship with the fern, namely the metabolites flow between the host and the symbionts, can be seen as a unique micro-ecosystem with own well-established characteristics. This association is maintained during the whole life cycle of the peteridophyte. The *Anabaena* apical colony is associated with shoot apex lacks heterocysts and, therefore, is unable to fix nitrogen. In mature leaves, the *Anabaena* filaments cease to grow and differentiate heterocysts, which are the site of N_2 fixation. Besides the cyanobacteria, a population of bacteria undergoes a pattern of infection identical to *Anabaena* and probably is the third partner of this symbiosis (Wallace and Gates 1986; Carrapico and Tavares 1989a; Carrapico 1991; Forni et al. 1989). The prokaryotic colony, cyanobacteria and bacteria, are also present in the sexual structures (sporocarps) of the fern (Carrapico 1991). The cyanobacterium is transferred from the sporophyte to the next generation via the megasporocarp. A cyanobacterium colony resides between the megasporocarp wall and the megasporangium one and inoculates the newly emerging sporophyte plant. A colony of the symbiotic cyanobacteria is formed near the shoot apex and thus, enables symbiosis to be established within the developing leaf cavities (Watanabe and Van Hove 1996). The presence of bacteria in the megasporocarps in association with the cyanobacteria also suggests a behavior pattern similar to the cyanobionts (Carrapico 1991). The presence of *Anabaena* throughout the life cycle of the fern favors the obligatory nature of the symbiosis and suggests a parallel phylogenetic evolution of both partners (Watanabe and Van Hove 1996).

This symbiotic association is the only fern-cyanobacteria association that presents agricultural interest by the nitrogen input that this plant can introduce in the fields; and for that reason, it has been used in several tropical and subtropical countries in different continents (Moore 1969; Kannayan 1986; Van Hove and Diara 1987; Shi and Hall 1988; Wagner 1997). Historically, *Azolla* has been used as green manure for wetland rice in northern Vietnam and central to southern China for centuries (Nierzwicki-Bauer 1990; Watanabe and Van Hove 1996).

Azolla filiculoides (Red *Azolla*) is the only member of this genus, and of the family Azollaceae common on farm dams, and other still-water bodies. The plants are small (usually only a few centimeters across) and flat, but can be very abundant and form large mats. The plants are typically red and have very small water repellent leaves. *Azolla* floats on the surface of water by means of numerous small and closely overlapping scale-like leaves, with their roots hanging in the water. They form a symbiotic relationship with the cyanobacterium *Anabaena azollae* which fixes atmospheric nitrogen, giving the plant access to the essential nutrients. This has led to the plant being named as a "super-plant", as it can readily colonize the areas of freshwater and grow at great speed while doubling its biomass in every two to three days. The only known limiting factor on its growth is phosphorus, another essential mineral. An abundance of phosphorus, due for example to eutrophication or chemical runoff, often leads to *Azolla* blooms.

The nitrogen-fixing capability of *Azolla* has led to its being widely used as a biofertilizer, especially in parts of South-east Asia. Indeed, the plant has been used to bolster agricultural productivity in China for over a thousand years. When rice paddies are flooded in the spring, they can be inoculated with *Azolla*, which then quickly multiplies to cover the water, suppressing weeds. The rotting plant material releases nitrogen to the rice plants, providing up to nine tones of protein per hectare per year (http://www.fao.org).

Azolla are also serious weeds in many parts of the world, entirely covering bodies of water at some places. The myth that no mosquito can penetrate the coating of fern to lay its eggs in the water gives the plant its common name "mosquito fern" (http://www.americaswetlandresources.com).

Most of the species can produce large amounts of deoxyanthocyanins in response to various stresses (Wagner 1997) including bright sunlight and extremes of temperature (Moore 1969; Zimmerman 1985) causing the water surface to appear to be covered with an intensely red carpet. Herbivore feeding induces accumulation of deoxyanthocyanins and leads to a reduction in the proportion of polyunsaturated fatty acids in the fronds, thus lowering their palatability and nutritive value (Cohen et. al. 2002).

Azolla cannot survive winters with prolonged freezing, so is often grown as an ornamental plant at high latitudes where it cannot establish itself firmly enough to become a weed. It is not tolerant to salinity; normal plants can't survive in greater than 1.0–1.6% and even conditioned organisms die in over 5.5% salinity (Brinkhuis et. al. 2006).

Azolla reproduces sexually and asexually by splitting. Like all ferns, sexual reproduction leads to spore formation, but *Azolla* sets itself apart from other members of its group by producing two kinds. During the summer months, numerous spherical structures called sporocarps form on the undersides of the branches. The male sporocarp is greenish or reddish and looks like the egg mass of an insect or spider. It is two millimeters in diameter and inside are numerous male sporangia. Male spores (microspores) are extremely small and are produced inside each microsporangium. Curiously, microspores tend to adhere in clumps called massulae (Arnold 1955).

Female sporocarps are much smaller, containing one sporangium and one functional spore. Since an individual female spore is considerably larger than a male spore, it is termed as a megaspore.

Azolla has microscopic male and female gametophytes that develop inside the male and female spores. The female gametophyte protrudes from the megaspore and bears a small number of archegonia, each containing a single egg. The microspore forms a male gametophyte with a single antheridium which produces eight swimming sperms (Robert et. al. 1965). The barbed glochidia on the male spore clusters are assumed to cause them to cling to the female megaspores, thus facilitating fertilization.

In addition to its traditional cultivation as a biofertilzer for wetland paddy (due to its ability to fix nitrogen), *Azolla* is finding increasing use for sustainable production of livestock feed (Pillai et al. 2008). *Azolla* is rich in proteins, essential amino acids, vitamins and minerals. Studies describe that feeding *Azolla* to dairy cattle, pigs, ducks and chickens resulted in reported increase in milk production, weight of broiler chickens and egg production of layers, as compared to conventional feed. One FAO study describes how *Azolla* integrates into a tropical biomass agricultural system, reducing the need for inputs (Preston and Murgueitio 2011). *Azolla* has also been suggested as food stuff for human consumption. However, no long term studies about the healthiness of eating *Azolla* have been made on humans (Sjodin Erik 2012).

Azolla has been used, for at least one thousand years (http://www.tropicos. org), in rice paddies as a companion plant, because of its ability to both fix nitrogen and block out light to prevent any competition from other plants, aside from the rice, which is planted when tall enough to poke out of the water through the *Azolla* layer. Mats of mature *Azolla* can also be used as weed-suppressing mulch.

As an additional benefit to its role as a paddy biofertilizer, *Azolla* spp. have been used to control mosquito larvae in rice fields. The plant grows in a thick mat on the surface of the water, making it more difficult for the larvae to reach the surface to breathe, effectively choking the larvae (Myer et. al. 2008).

A study of Arctic Paleoclimatology reported that *Azolla* may have had a significant role in reversing an increase in greenhouse effect that occurred 55 million years ago that caused the region around the North Pole to turn into a hot tropical environment. This research conducted by the Institute of Environmental Biology at Utrecht University claims that massive patches of *Azolla* growing on the (then) freshwater surface of the Arctic Ocean consumed enough carbon dioxide from the atmosphere for the global greenhouse effect to decline, eventually causing the formation of ice sheets in Antarctic and the current "Icehouse period" which we are still in. This theory has been termed as the *Azolla* event.

Azolla is used as biofertilizer in low-land rice ecosystem and found to contribute 40–60 kg of N/ha per rice crop. The use of *Azolla* as dual crop with rice is more significant in South-East Asian countries. *Azolla* biomass enriches the soil organic matter in addition to its N contribution.

Anabaena Azolla and endosymbiont in symbiotic association with different species of *Azolla* fixes atmospheric nitrogen. The endosymbiont is associated with the dorsal lobe of the *Azolla* fronds. *Azolla* provides required carbon to the endosymbiont, which fixes nitrogen and transfers to *Azolla*. The heterocyst of Anabaena is the site of nitrogen fixation, while the vegetative cells are involved in photosynthesis. The production of ammonia by the algal symbiont is partly utilized by *Azolla* and the remained is excreted into the surrounding system.

The species of *Azolla*, *A.microphylla*, is tolerant to high temperature up to 38°C and well-adopted to rice field conditions of tropical climate. The doubling time is 2–3 days; however, the environmental conditions largely influence the growth and multiplication. The presence of *Anabaena azollae* is associated with only megasporocarp. *Anabaena azollae* has three types of cells viz., vegetative cells, i.e. the site of photosynthesis, heterocysts, i.e. the site of nitrogen fixation and akinates, i.e. the thick-walled resting spore. *Azolla* can be grown in the N-free medium.

(2)   Non-symbiotic N_2 fixing biofertllizers

These biofertlizers fix atmospheric N_2 freely in the rhizosphere without having any symbiotic association with the crop plants. Such non-symbiotic diazotrophic biofertilizers are *Azotobacter*, *Beijerinckia*, and blue-green algae, etc.

Azotobacter

The *Azotobacter* genus was discovered in 1901 by Dutch microbiologist and botanist Martinus Beijerinck, who was one of the founders of environmental microbiology. He selected and described the species *Azotobacter* chroococcum – the first aerobic, free-living nitrogen fixer (Beijerinck 1901).

Azotobacter is a genus of usually motile, oval, or spherical bacteria that form thick-walled cysts and may produce large quantities of capsular slime. They are aerobic, free-living soil microbes which play an important role in the nitrogen cycle in nature, binding atmospheric nitrogen, which is inaccessible to plants and releasing it in the form of ammonium ions into the soil. Apart from being a model organism, it is used by humans for the production of biofertilizers, food additives and few biopolymers. *Azotobacter* species are gram-negative bacteria found in neutral and alkaline soils (Gandora et. al. 1998; Martyniuk and Martyniuk 2003) in water and in association with few plants (Tejera et. al. 2005; Kumar et. al. 2007).

Azotobacter species are ubiquitous in neutral and weakly basic soils, but not acidic soils (Yamagata and Itano 1923). They are also found in the Arctic and Antarctic soils, despite the cold climate, short growing season and relatively low pH values of these soils (Boyd and Boyd 1962). In dry soils, *Azotobacter* can survive in the form of cysts for up to 24 years (Moreno et. al. 1986).

Representatives of the genus *Azotobacter* are also found in aquatic habitats, including freshwater (Johnstone 1967) and brackish marshes (Dicker and

Smith 1980). Several members are associated with plants and are found in the rhizosphere, having certain relationships with the plant (Berkum and Bohlool 1980). Some strains are also found in the cocoons of the earthworm *Eisenia fetida* (Zachmann and Molina 1993).

Cells of the genus *Azotbacter* are relatively large for bacteria (1–2 µm in diameter). They are usually oval, but may take various forms from rods to spheres. In microscopic preparations, the cells can be dispersed or form irregular clusters or occasionally chains of varying lengths. In fresh cultures, cells are mobile due to the numerous flagella (Baillie et. al. 1962). Later, the cells lose their mobility, become almost spherical and produce a thick layer of mucus, forming the cell capsule. The shape of the cell is affected by the amino acid glycine which is present in the nutrient medium peptone (Vela and Rosenthal 1972).

The cells show inclusions under magnification, some of which are colored. In the early 1900s, the colored inclusions were regarded as "reproductive grains", or gonidia – a kind of embryo cell (Jones 1920). However, it was later demonstrated that the granules do not participate in the cell division (Lewis 1941). The colored grains are composed of voutin; whereas the colorless inclusions are drops of fat, which act as energy reserves (Lewis 1937).

Cyst of the genus *Azotobacter* are more resistant to adverse environmental factors than the vegetative cells; in particular, they are twice more resistant to UV light. They are also resistant to drying, ultrasound, gamma, and solar irradiation, but not to heating (Socolofsky and Wyss 1962).

The formation of cysts is induced by the changes in the concentration of nutrients in the medium and addition of some organic substances such as ethanol, n-butanol, or ß-hydroxybutyrate. Cysts are rarely formed in liquid media (Layne and Johnson 1964). The formation of cysts is induced by chemical factors and is accompanied by metabolic shifts, changes in catabolism, respiration and biosynthesis of macromolecules (Sadoff 1975). It is also affected by aldehyde dehydrogenase (Gama-Castro et. al. 2001) and the response regulator AlgR (Nunez et. al. 1999).

The cysts of *Azotobacter* are spherical and consist of the so-called 'central body' – a reduced copy of vegetative cells with several vacuoles- and the 'two-layer shell'. The inner part of the shell is called intine and has a fibrous structure (Pope and Wyss 1970). The outer part has a hexagonal crystalline structure and is called exine (Page and Sadoff 1975). Exine is partially hydrolyzed by trypsin and is resistant to lysozyme, in contrast to the central body (Lin and Sadoff 1969). The central body can be isolated in a viable state by some chelation agents (Parker and Socolofsky 1968). The main constituents of the outer shell are alkylresorcinols that are composed of long aliphatic chains and aromatic rings. Alkylresorcinols are also found in other bacteria, animals, and plants (Funa et. al. 2006).

A cyst of the genus *Azotobacter* is the resting form of a vegetative cell; whereas, usual vegetative cells are reproductive. The cyst of *Azotobacter* does not serve this purpose and is necessary for surviving adverse environmental factors. Following the resumption of optimal environmental conditions, which include a certain value of pH, temperature and source of carbon, the cysts germinate and the newly formed vegetative cells multiply by a simple division. During the germination, the cysts sustain damage and release a large vegetative cell. Microscopically, the first manifestation of spore germination is the gradual decrease in light refractive by cysts, which is detected with phase contrast microscopy. Germination of cysts, a slow process, takes about 4–6 hours. During germination, the central body grows and captures the granules of volutin, which are located in the intima (the innermost layer). Then the exine bursts and the vegetative cell is freed from the exine, which has a characteristic horseshoe shape (Wyss et. al. 1961). This process is accompanied by metabolic changes. Immediately after being supplied with a carbon source, the cysts begin to absorb oxygen and emit carbon dioxide; the rate of this process gradually increases and saturates after four hours. The synthesis of proteins and RNA occurs in parallel, but it intensifies only after five hours after the addition of the carbon source. The synthesis of DNA and nitrogen fixation are initiated 5 ha after the addition of glucose to a nitrogen–free nutrient medium (Loperfido and Sadoff 1973).

Germination of cysts is accompanied by changes in the intima, visible with an electron microscope. The intima consists of carbohydrates, lipids and proteins and has almost the same volume as the central body. During germination of cysts, the intima hydrolyses and is used by the cell for the synthesis of its components (Lin et. al. 1978).

Azotobacter respires aerobically while receiving energy from redox reactions, using organic compounds as electron donors. *Azotobacter* can use a variety of carbohydrates, alcohols, and salts of organic acids as sources of carbon and can fix at least 10 μg of nitrogen per gram of glucose consumed. Nitrogen fixation requires molybdenum ions, but they can be partially replaced by vanadium ions or even omitted altogether. The source of nitrogen can be nitrates, ammonium ions, or aminoacids. The optimal pH for the growth and nitrogen fixation is 7.0–7.5, but growth is sustained in the pH range from 4.8 to 8.5 (George and Garrity 2005). *Azotobacter* can also grow mixotrophically, in a nitrogen-free medium containing mannose; this growth mode is hydrogen-dependent. Hydrogen is available in the soil; thus, this growth mode may occur in nature (Wong and Maier 1985).

While growing, *Azotobacter* produces flat, slimy, paste-like colonies with a diameter of 5–10 mm, which may form films in liquid nutrient media. The colonies can be dark-brown, green, or of some other colors, or may be colorless, depending on the species. The growth is favored at a temperature of 20–30°C (Tepper et. al. 1979).

Bacteria of the genus *Azotobacter* are also known to form intracellular inclusions of polyhydroxyalkanoates under certain environmental conditions (e.g., lack of elements such as phosphorus, nitrogen, or oxygen combined with an excessive supply of carbon sources).

Azotobacter produces pigments. For example, *Azotobacter chroococcum* forms a dark-brown water-soluble pigment melanin. This process occurs at high levels of metabolism during the fixation of nitrogen and is thought to protect the nitrogenase system from oxygen (Shivprasad and Page 1989). Other *Azotobacter* species produce pigments from yellow-green to purple colors (Jensen 1954), including a green pigment which fluoresces with a yellow-green light and a pigment with blue-white fluorescence.

Azotobacter species are free-living nitrogen-fixing bacteria; in contrast to *rhizobium* species, they normally fix molecular nitrogen from the atmosphere without symbiotic relations with plants, although some *Azotobacter* species are associated with plants (Kass et. al. 1971). Nitrogen fixation is inhibited in the presence of available nitrogen sources, such as ammonium ions and nitrates (Burgmann et. al. 2003).

Azotobacter species have a full range of enzymes needed to perform the nitrogen fixation: ferredoxin, hydrogenases and an important enzyme nitrogenase. The process of nitrogen fixation requires an influx of energy in the form of adenosine triphosphate. Nitrogen fixation is highly sensitive to the presence of oxygen, so *Azotobacter* developed a special defensive mechanism against oxygen, namely significant intensification of metabolism that reduces the concentration of oxygen in the cells (Shank et. al. 2005). Also, a special nitrogenase-protective protein protects nitrogenase and is involved in protecting the cells from oxygen. Mutants not producing this protein are killed by oxygen during nitrogen fixation in the absence of a nitrogen source in the medium (Maier and Moshiri 2000). Homocitrate ions play a certain role in the processes of nitrogen fixation by *Azotobacter* (Durrant et. al. 2006).

Nitrogenase is the most important enzyme involved in nitrogen fixation. *Azotobacter* species have several types of nitrogenase. The basic one is molybdenum-iron nitrogenase (Howard and Rees 2006). An alternative type contains vanadium; it is independent of molybdenum ions (Bellenger et. al. 2008; Ruttimann-Johnson et. al. 2003; Robson et. al. 1986) and is more active than the Mo-Fe nitrogenase at low temperatures. So it can fix nitrogen at temperatures as low as 5°C and its low-temperature activity is 10 times higher than that of Mo-Fe nitrogenase (Miller and Eady 1988). An important role in maturation of Mo-Fe nitrogenase plays the so-called P-cluster (Hu et. al. 2007). Synthesis of nitrogenase is controlled by the nif genes (Curatti et. al. 2005). Nitrogen fixation is regulated by the enhancer protein Nif A and the "sensor" flavoprotein Nif L which modulates the activation of gene transcription of nitrogen fixation by redox-dependent switching (Hill et. al. 1996). This regulatory mechanism, relying on two proteins forming complexes with each other, is uncommon for other systems (Money et. al. 2001).

Nitrogen fixation plays an important role in the nitrogen cycle. *Azotobacter* also synthesizes some biologically active substances, including some phytohormones such as auxins (Ahmad et. al. 2005); thereby stimulating plant growth (Rajee et. al. 2007). They also facilitate the mobility of heavy metals in the soil, thus enhancing bioremediation of soil from heavy metals, such as cadmium, mercury and lead (Chen et. al. 1995). Some kinds of *Azotobacter* can also biodegrade chlorine-containing aromatic compounds, such as 2,4,6-trichlorophenol. The latter was previously used as an insecticide, fungicide and herbicide, but later was found to have mutagenic and carcinogenic effects (Li et. al. 1991).

Owing to their ability to fix molecular nitrogen and by this increasing the soil fertility and stimulating plant growth, *Azotobacter* species are widely used in agriculture (Neeru 2000) particularly in nitrogen biofertilizers such as azotobacterin. They are also used in production of alginic acid (Galindo et. al. 2007; Page et. al. 2001; Ahmed and Ahmed 2007), which is applied in medicine as an antacid in the food industry as an additive to ice cream, puddings and creams (Hans Gunter Schlegel et. al. 1993) and in the biosorption of metals (Emitiazia et. al. 2004).

The species of *Azotobacter* are *Azotobacter chroococcum, Azotobacter paspali, Azotobacter agilis, Azotobacter armeniacus, Azotobacter beijerinckii, Azotobacter nigricans, Azotobacter salinestris, Azotobacter tropicalis,* and *Azotobacter vinelandii.* Among these, *Azotobacter chroococcum* is widely used for the production of bioferilizers.

The *Azotobacter chroococcum* are large ovoid rods shaped cell of 2 × 5 μ in size. Occurs frequently in pairs and motile with peritrichous flagella. Cysts and capsular slime are formed. It is gram negative and aerobic. Optimum temperature is between 20 and 30°C and pH 7.0–7.5, as required for growth. No water-soluble pigment is produced; but growth on agar media is characterized by a water-insoluble brown pigment, which later becomes black. Utilizes starch is unique among the species of the genus. Mannitol is utilized, but not rhamnose. The G + C content of DNA ranges from 65–66 moles %.

Habitat: The cells are found in both soil and water.

Beijerinckia

The existence of a non-symbiotic relationship between grasses and micro-organisms was proposed by Parker (1957) and postulated that nitrogen was fixed in the rhizosphere by micro-organisms utililizing root excretions. Dobereiner (1961, 1966) observed that large number of nitrogen-fixing bacteria *Beijerinckia indica* occurred in the rhizosphere and rhizoplane of sugarcane.

In contrast to the ubiquitous distribution of the nitrogen-fixing bacteria of the genus *Azotobacter* which can be encounter in all soils with suitable pH, the nitrogen-fixing bacteria of the genus Beijerinckia seem to be restricted to the tropics. All attempts to isolate Beijerinckia from north-european soils have so far failed. The presence of Beijerinckia is almost invariably connected with the occurrence of laterites, tropical-red earths on latosols. It was also suggested that Beijerinkia might occur in lateritic soil-fossils or developed under particular climatic conditions outside the tropics. Becking (1959) gives first evidence of the occurrence of Beijerinckia in lateritic soils outside the tropics in South Africa.

Soil temperature, soil moisture, light, soil pH, nitrogenous fertilizer, C/N ratio and ease of decomposition of organic matter affect the ability of the grass-bacteria association to fix nitrogen. Acetylene reduction values equivalent to nitrogen fixation rates up to 100 kg N/hact/year have been estimated (Weier 1980).

Beijerinckia are free living, unicellular, aerobic, chemoheterotrophic bacteria with the ability to fix atmospheric nitrogen. They can be distinguished from other nitrogen fixing bacteria by cell morphology and some physiological characteristics. Members of this genus have typical rod-shaped cells when young; pear, or dumbbell-shaped cells with round ends containing polar lipoid bodies after aging. Beijerinckia species show great acid tolerance, being able to grow and fix nitrogen at pH 3.0–4.0. These bacteria are extremely acid tolerant and have been shown to fix more nitrogen as the acidity levels increase (down to 2.7 pH) (Barbosa et. al. 2002). The microbe is more prominent in acidic soils, yet is often found in alkali soils as well. This type of species is Beijerinckia indica (Derx 1950).

Being found in almost every soil on Earth, this bacteria plays important roles in plant-growth due to its metabolism. It can survive cold weather, making it ideal for temperate regions that experience a seasonal freeze. It produces a slime that makes it easier for water retention in rainforest soils. It also is a provider of fixed N_2 from the atmosphere. The pigments it secretes may be important in the stimulation of other microbes in the surrounding environment.

The cell of Beijerinckia generally ranges from 1.7 to 4.5 µm in length and about 0.5–1.5 µm in diameter. Beijerinckia derxii is generally smaller than *Azotobacter*. The temperature range is from 10° to 35°C. They contain two lipoid structures at the polar ends of the cell. These structures are highly refractive and may be involved with light protection. Theses lipoids consist of poly-ß-hydroxybutyrate (PHB).

It has the ability to change a liquid media viscous by polysaccharide slime production. On surface, the media they produce raises colonies of highly elastic slime in varying colors. B. derxii can be characterized by a green fluorescent pigment; this pigment is most prominent on iron-deficient media. The exopolysaccharides may also protect the nitrogenase from oxygen damage, as it forms a protective O_2 barrier.

Low O_2 availability increases the production of nitrogenase; similar increases were seen in the presence of thoisulfate and in a low pH (2.7) (Barbosa et. al. 2002). Similarly, low levels of carbohydrates produce the same effect.

Carbon utilized includes a large array of sugars, organic alcohols and organic acids. They can also hydrolyze starches. They have been shown in alkali medium to decrease the pH to about 4.0–5.0; this is accomplished by acetic acid and some lactic acid production (Becking 1961). They require molybdenum in their nitrogen fixation and for optimal growth. $CaCO_3$, a necessary mineral in most *Azotobcater* species, inhibits the lag phase of growth in Beijerinckia.

It was initially isolated from a Malaysian quartzite soil by Altson in 1936; and again in Australia by Tchan in 1957. It can withstand very acidic conditions. This microbe has also been isolated from Europe, South America, Continental Asia, China, Japan, Australia and Pacific North-west soils in the America, suggesting that the microbe has worldwide distribution. It has been shown to resist freezing and the strands are viable indefinitely when frozen in liquid nitrogen. No reduction of viability occurs when stored for 3–4 months at 4°C (Becking 1961).

It has also been studied that how B. derxii is able to stimulate plant growth as well as other plant responses with releasing certain plant-growth regulators, specifically indole acetic acid (IAA), ethylene, polyamines, and certain amino acids into the soil (Thuler et. al. 2003).

It has been shown that within the Rhizosphere of rice plants, Beijerinckia numbers increase. No evaluation has yet been made to this phenomenon. Similar occurrences have been reported in sugar cane soils. One experiment demonstrated that sorghum strongly stimulates the growth of Beijerinckia. Apart from being within soil and water habitats, Beijerinckia have also been isolated within both the rhizosphere and the phyllosphere of plants.

Beijerinckia fix atmospheric nitrogen non-symbiotically in acidic soils of high rainfall area. The species of Beijerinckia *are Beijerinckia indica, Beijerinckia mobilis, Beijerinckia derxii,*and *Beijerinckia fluminensis.* The characters of *Beijerinckia indica* are: single cells, straight, slightly curved or pear rods of 0.5–1.2 μ × 1.6–3.0 μ size, non-motile or motile with peritrichous flagella. Molybdenum is required for nitrogen fixation, but calcium is not required for growth and nitrogen fixation. Utilize glucose, sucrose and fructose as a carbon source. On Agar medium, colonies of Beijerinckia are raised, semi-transparent which soon become uniformly turbid or opaque white. On aging, colonies develop a light-reddish, pink, cinnamon, or fawn color on neutral or alkaline media. On acidic media, colonies remain colorless. On acidic media, the slime produced is more tenacious, tough and elastic than on alkaline media. Giant colonies may develop first with a smooth surface, which later on turn into folded, wrinkled, or plicate surface. Liquid media become viscous by slime production; highly acidic tolerant, optimum pH 4–10, and temperature range about 10–35°C is required for growth.

Habitat: It is mostly found in acidic soil.

Blue-green algae

Blue-green algae or cyanobacteria, also known as cyanophyta, are a phylum of bacteria which obtain their energy through photosynthesis (http://www.ucmp.berkeley.edu/bacteria/cyanolh.html). The name "cyanobacteria" comes from the color of the bacteria. They are often called blue-green algae, but some consider that name a misnomer as cyanobacteria are prokaryotic and algae should be eukaryotic (Allaby 1992), although other definitions of algae encompass prokaryotic organisms (Lee 2008).

Cyanobacteria can be found in almost every terrestrial and aquatic habitat like oceans, fresh water, damp soil, temporarily moistened rocks in deserts, bare rock and soil and even Antarctic rocks.

They can occur as planktonic cells or form phototrophic biofilms. They are found in almost every endolithic ecosystem (De los Rios 2007). A few are endosymbionts in lichens, plants, various protists, or sponges and provide energy for the host. Some live in the fur of sloths, providing a form of camouflage (Vaughan Terry 2011).

Aquatic cyanobacteria are known for their extensive and highly visible blooms which can form in both freshwater and marine environments. The blooms can have the appearance of blue-green paint or scum. These blooms can be toxic, and frequently lead to the closure of recreational waters when spotted. Marine bacteriophages are significant parasites of unicellular marine cyanobacteria (Nora Schultz 2009).

Cyanobacteria are photosynthetic nitrogen-fixing group that survive in a wide variety of habitats, soil and water. In this group, photosynthetic pigments are cyanophyscyne, allo-phycocyanine and erythro-phycocyanine. Their thallu varies from unicellular to filamentous, heterocystous. They fix atmospheric nitrogen in aerobic condition by heterocyst, specialized cell, and in anaerobic condition.

Many cyanobacteria form motile filaments of cells called hormogonia that travel away from the main biomass to bud and form new colonies elsewhere. The cells in a hormogonium are often thinner than in the vegetative state, and the cells on either end of the motile chain may be tapered. In order to break away from the parent colony, a hormogonium often must tear apart a weaker cell in a filament called a necridium.

Each individual cell of a cyanobacterium typically has a thick and gelatinous cell wall. They lack flagella, but hormogonia of few species can move about by gliding along surfaces. Many of the multi-cellular filamentous forms of *Oscillatoria* are capable of a waving motion; the filament oscillates back and forth. In water columns, some cyanobacteria float by forming gas vesicles as in archaea. These vesicles are not organelles as such. They are not bounded by lipid membranes but by a protein sheath.

Some of these organisms contribute significantly to global ecology and the oxygen cycle. The tiny marine cyanobacterium *Prochlorococcus* was discovered in 1986 and accounts for more than half of the photosynthesis of the open ocean (Steve Nadis 2003). Many cyanobacteria even display the circadian rhythms that were once thought to exist only in eukaryotic cells.

As with any prokaryotic organism, cyanobacteria do not have nuclei or an internal membrane system. However, many species of cyanobacteria have folds on their external membranes that function in photosynthesis. Cyanobacteria get their color from the bluish pigment phycocyanin, which they use to capture light for photosynthesis. In general, photosynthesis in cyanobacteria uses water as an electron donor and produces oxygen as a by-product; though some may also use hydrogen sulfide (Cohen et. al. 1986) – a process which occurs among other photosynthetic bacteria such as the purple sulfur bacteria. Carbon dioxide is reduced to form carbohydrates via the Calvin cycle. In most forms, the photosynthetic machinery is embedded into folds of the cell membranes known as thylakoids. The large amount of oxygen in the atmosphere is considered to have been first created by the activities of ancient cyanobacteria. They are often found as symbionts with a number of other groups of organisms such as fungi (lichens), corals, pteridophytes (*azolla*), angiosperms (gunnera), etc.

Many cyanobacteria are able to reduce nitrogen and carbon dioxide under aerobic conditions, a fact that may be responsible for their evolutionary and ecological success. The water-oxidizing photosynthesis is accomplished by coupling the activity of photosystem (PS) II and I (Z-scheme). In anaerobic conditions, they are also able to use only PS I-cyclic photophosphorylation with electron donors other than water (hydrogen sulfide, thiosulphate, or even molecular hydrogen just like purple photosynthetic bacteria. Furthermore, they share an archaeal property and the ability to reduce elemental sulfur by anaerobic respiration in the dark. Their photosynthetic electron transport shares the same compartment as the components of respiratory electron transport. Their plasma membrane contains only the components of the respiratory chain, while the thylakoid membrane hosts an interlinked respiratory and photosynthetic electron transport chain. The terminal oxidases in the thylakoid membrane respiratory/photosynthetic electron transport chain are essential for survival to rapid light changes, although not for dark maintenance under conditions where cells are not light stressed (Lea-Smith et. al. 2013).

Attached to thylakoid membrane, phycobilisomes act as light harvesting antennae for the photsystems. The phycobilisome components (phycobiliproteins) are responsible for the blue-green pigmentation of most cyanobacteria. The variations on this theme are mainly due to carotenoids and phycoerythrins which give the cells the red-brownish coloration. In some cyanobacteria, the color of light influences the composition of phycobilisomes. In green light, the cells accumulate more phycoerythrin; whereas in red light they produce more phycocyanin. Thus, the bacteria appear green in red light and red in green light. This process is known as complementary chromatic adaption and is a way for the cells to maximize the use of available light for photosynthesis.

A few genera, however, lack phycobilisomes and have chlorophyll b instead (*Prochloron*, *Prochlorococcus*, *Prochlorothrix*). These were originally grouped together as the prochlorophytes or chloroxybacteria, but appear to have developed in several different lines of cyanobacteria. For this reason, they are now considered as part of the cyanobacterial group.

Cyanobacteria use the energy of sunlight to drive photosynthesis, a process where the energy of light is used to split water molecules into oxygen, protons, and electrons. While most of the high-energy electrons derived from water are used by the cyanobacterial cells for their own needs, a fraction of these electrons are donated to the external environment via electrogenic activity (Pisciotta et. al. 2010). Cyanobacterial electrogenic activity is an important microbiological conduit of solar energy into the biosphere.

Cyanobacteria include unicellular and colonial species. Colonies may form filaments, sheets, or even hollow balls. Some filamentous colonies show the ability to differentiate into several different cell types: vegetative cells, the normal and photosynthetic cells that are formed under favorable growing conditions; akinetes, the climate resistant spores that may form when environmental conditions become harsh; and thick-walled heterocysts, which contain the enzyme nitrogenase, vital for nitrogen fixation. Heterocysts may also form under the appropriate environmental conditions (anoxic) when fixed nitrogen is scarce. Heterocyst-forming species are specialized for nitrogen fixation and are able to fix nitrogen gas into ammonia (NH_3), nitrites (NO_2), or nitrates (NO_3), which can be absorbed by plants and converted to protein and nucleic acids (atmospheric nitrogen is not bio-available to plants).

Rice plantations utilize healthy population of nitrogen-fixing cyanobacteria (Anabaena, as symbiotes of the aquatic fern azolla) for use as rice paddy fertilizer (Stefano Bocchi and Antonino Malgioglio 2010).

Cyanobacteria are arguably the most successful group of microorganisms on earth. They are the most genetically divers species; they occupy a broad range of habitats across all latitudes, widespread in freshwater, marine and terrestrial ecosystems, and they are found in the most extreme niches such as hot springs, salt works and hypersaline bays. Photoautotrophic, oxygen-producing cyanobacteria, created the conditions in the planet's early atmosphere that directed the evolution of aerobic metabolism and eukarotic photosynthesis. Cyanobacteria fulfill vital ecological functions in the world's oceans, being important contributors to global carbon and nitrogen budgets (Stewart and Falconer 2008).

Algae are a heterogeneous assemblage of holophytic plants, which have failed to reach the specialization of archegoniate plants. This is the only group which includes both prokaryotic and eukaryotic organisms. These are the pioneer colonizers and also occupy the base of the trophic pyramid. Out of the total annual addition of 2×10^{11} tonnes of organic matter on this planet, as much as 0.8×10^{11} tonnes is contributed by the algae. They are cosmopolitan as they are able to withstand extremes of temperature and moisture stress because of the presence of a higher proportion of main valency bonds in their proteins, and the protection provided by the polysaccharide envelope around their trichomes.

Many blue-green algae have the specialized cells known as heterocysts. These modified vegetative cells because of their thick walls and absence of pigment system II, have active nitrogenase enzymes and act as sites for nitrogen fixation. Although nitrogenase is present in the vegetative cells also, it remains inactive because of the liberation of oxygen during photosynthesis. Their trophic independence for carbon and nitrogen makes them responsible for the natural fertility build-up of the rice field soils.

The property of converting molecular nitrogen to ammonia was considered to be the morphology of the heterocystous blue-green algae like *Anabaena*, *Nostoc, Cylindrospermum, Tolypothrix. Calothri, Scytonema, Hapalosiphon, Westiellopsisi*, and *Stigonema* belonging to the orders Nostocales and Stigonematales. Now about 12 non-heterocystous cyanobacteria are known to utilize atmospheric nitrogen. While most of them need anaerobic or microaerophilic conditions, forms like *Gloeocapsa* fix nitrogen under aerobic conditions.

The diazotrophic cyanobacteria form an inexpensive farm-grown input, which help in better crop nutrient management while working in perfect harmony with nature. They benefit the crop by providing the biologically fixed nitrogen. In addition, they synthesize and liberate growth-promoting substances like vitamins and auxins. Being photolithotrophic in nature, they contribute organic matter in the soil. The process of photosynthesis provides oxygen to the submerged roots of rice plants and also helps in reducing oxidisable matter content in the soil. The mucilaginous sheath around the trichomes binds the soil particles and improves the soil aggregation. Increased size of the soil aggregates improves the water holding capacity and aeration of the soil.

Many algae have been shown to solubilize the insoluble phosphates. This gains importance in view of the fact that most of the phosphatic fertilizers when applied to the soil, immediately convert into calcium phosphate and become unavailable to the plants. These algae are resistant to the conventionally used pesticides and other agro-chemicals. They have also been shown to have ameliorating effect on the saline-alkaline soils.

High temperature (30–40°C) and humidity, diffused light and pH range of 6.5–8.5 are the favorable conditions for the multiplication of the blue-green algae. These conditions prevail throughout the growing season of the rice crop. Although the algae grow on their own in the rice fields, but all of them are not useful for the crop. The green algae like *Spirogyra*, *Rhinoclonium*, *Hydrodictyon* and *Cladophora* complete the crop for nutrients and reduce tillering in rice plants. It is therefore, necessary to introduce region specific efficient strains of algae for better ecological benefits.

The significance of cyanobacteria as biofertilizer lies in the fact that unlike the chemical fertilizers, the crop does not utilize them themselves. Only the crop uses the products from their activities. During the crop-growing season, the blue-green algae grow and multiply, fix atmospheric nitrogen and make a part of it available to the crop. During the dry season they pernnate and become active with the onset of the favorable season. Thus, it is expected that superimposed algalization of the rice crop may increase their population in the soil to a level which will sustain the effect in the absence of fresh application.

The chemical fertilizers provide the nutrients in a cyclic manner, but the blue-green algae do so in a gradual and linear manner. This reduces the loss and helps the crop to utilize more of the applied nutrients. Algal application has been shown to increase the crop yield by 25–30%. The effect of algal application is more pronounced at lower levels of fertilizer nitrogen. In the presence of higher levels of fertilizer nitrogen, the supplementation effect of algalization is up to

25%. Algal inoculation provides approximate of up to 30 kg N/ha/season in the absence of fertilizer nitrogen.

In the presence of higher levels of fertilizer nitrogen, the algal contribution is supplemental. The concentration of combined nitrogen inhibiting growth is much higher than for nitrogen fixation. Also, the diazotrophic cyanobacteria appear to be more compatible with nitrate nitrogen than ammonium nitrogen. In the presence of combined nitrogen, the blue-green algae contribute to grow and show concentration dependent increase in the uptake of the added combined nitrogen.

Unlike the bacterial nitrogenase, algal nitrogenase has a "switch on" mechanism which is activated when the level of combined nitrogen falls below a threshold value. This enables the increased algal biomass to contribute more biologically fixed nitrogen when the level of fertilizer nitrogen in the ecosystem is reduced due to progressive utilization and loss.

Acetobacter

Dobereiner (1959) studied the association of nitrogen fixing bacteria with sugarcane by comparing inter row soil and rhizosphere soil samples, and reported that rhizosphere soil had more of nitrogen fixing organisms. Dobereiner (1961) further reported that 95% of the rhizosphere soil samples had *Beijerinkia* spp. in contrast to only 60% in the inter row soil. Sugarcane roots produced ethylene from acetylene up to 5 mol/g/h, showing nitrogenase activity (Dobereiner et. al. 1972). Ruschel et. al. (1975) maintained seedlings of sugarcane grown in compost/soil mixture in a 15 N atmosphere for 30 h and obtained the first evidence of nitrogen fixation.

Cavalcante and Dobereiner (1988) reported a new endophytic nitrogen fixing bacterium, occurring in large numbers in the xylem sap of sugarcane root, stem and leaf and proposed a new genus and species, viz. *Saccharobacter nitrocaptans*. Later, taxonomic studies by Gillis and De Ley (1980) showed that this bacterium belonged to the *Acetobacter* RNA cystron and confirmed it to be a new species under the genus *Acetobacter* by DNA/ DNA binding experiments. So, the nomenclature originally proposed as *Saccharobacter nitrocaptans* was changed to *Acetobacter diazotrophicus*.

Stephan et. al. (1991) reported that *Acetobacter* diazotrophicus fixed nitrogen even in the presence of Potassium nitrate at 10 mM and at low pH values (pH less than 3.0). These bacteria excreted almost half of the nitrogen fixed in form potentially available to the plants. Nitrogen fixing *Acetobacter* diazotrophicus, *Herbaspiririllum seropedicae* and *Herbaspiririllum rubrisubalbican* colonies are present in sugarcane plant and are capable of supplying high level of fixed nitrogen to this plant.

Many field and pot culture experiments with the bio-inoculant *Acetobacter* diazotrophicus on several varieties of sugarcane indicated a significant beneficial effect on the crop; not only in terms of nitrogen gain but also in increasing the yield of sugarcane (Muthukumarswamy et. al. 2002).

Acetobacter diazotrophicus are as follows:

(1) A gram negative, non-spore forming, microaerophilic diazotrophic bacterium.

(2) It is a straight rod with round ends and measures 0.7–0.9 u in length.

(3) Cells are motile with either a single or three lateral flagella.

(4) It has an optimum growth temperature of 30°C.

(5) The optimum pH for the bacterium is 4.0–4.5.

(6) On LGI medium, the colonies of *Acetobacter* diazotrophicus appear irregular, 2.3 mm in diameter, smooth, flat with bright yellow or yellow orange color.

(7) On potato dextrose agar medium (10% glucose), the colonies would produce dark-brown pigment.

(8) It requires a high concentration of sucrose or glucose (100–300 g/l) for growth and nitrogen assimilation.

(9) More importantly, this bacterium can fix nitrogen even in the presence of 25 mM of ammonium or 80 mM of nitrate in medium; this property of the bacterium makes it distinctly different from other diazotrophs whose nitrogenase enzyme succumbs to even low levels of combined nitrogen.

(10) *Acetobacter* diazotrophicus is more sensitive to antibiotics like ampicilin and acetic tetracycline and fairly resistant to streptomycin.

(11) It survives poorly in soil.

The bacterium is reported from all major sugarcane growing areas of Brazil, Australia, Mexico, Cuba, and India (Cavalcante and Dobereiner 1988; Muthukumarswamy et. al. 2000).

Endophytic nature of *A. diazotrophicus* was reported by Cavalcante and Dobereiner et. al. (1988). The bacterium infects sugarcane through damaged tissue (Boddey and Dobereiner 1994). Many authors described that *A. dizaotrophicus* is spread among cane cultivars by the mealy bugs associated with sugarcane as well as by the spores of the vesicular arbuscular mycorrhizal fungus (Paula et. al. 1992). They also observed that VAM fungi increases the translocation of *A. diazotrophicus*. James et. al. (1994) reported that *A. dizaotrophicus* first colonizes the root and lower stem epidermal surface of sugarcane during the first 15 days after planting and then uses root tips and lateral root junctions to enter the sugarcane plants where it was distributed around the plant in the translocation stream.

Beneficial effects of using bacterial inoculants include the production of plant growth promoting substances such as gibberellins (Lee et. al. 1970; Brown 1972). Calvalcante and Dobereiner (1988) reported that *A. diazotrophicus* promote root development and improve sugarcane growth. Libbert and Risch (1969) showed that epiphytic bacteria could increase the IAA content in the plant. Fuentez-Ramirez et. al. (1993) showed that all the strains of *A. diazotrophicus* produced indole acetic acid in defined supplemented with tryptophan and the amount of IAA produced was estimated from 0.14 to 2.42 ug/ml in culture medium.

Muthukumarswamy et. al. (2000) reported that *A. diazotrophicus* increase cane yield by 7–10 tonnes/acre with 50% reduction in the recommended chemical nitrogen and also observed an increase in sugar recovery from 0.5 to 1%. *A. diazotrophicus* combination with 50% nitrogen recorded cane yield and sugar yield values on par with 100% nitrogen applied treatments. They also reported that the growth and cane yield was better for the *Acetobacter* and *Azotobacter* and combination of *Acetobacter* and *Azotobacter* was better than other biofertilizer combinations.

Habitat: *Acetobacter* diazotrophicus occurs in the roots, stems, leaves (Calvalcante and Dobereiner 1988; Gillis et. al. 1989), rhizosphere soil of the sugarcane and even in cane juice (Muthukumarswamy et. al. 2000) in appreciable number. In addition, it could be isolated from cameroon grass (*Pennisetum purpureum* cv. *cameroon.*) and sweet potato (Paula et. al. 1992) as well as from different genera of mealy bugs associated with sugarcane plants (Boddey and Dobereiner 1994).

(3)   *Associative nitrogen fixing biofertilizers*

Azospirillum

The genus *Azospirillum* was first described by Tarrand et. al. (1978) with two species, *Azospirillum lipoferum* and *Azospirillum brasilense*. At present the genus comprises of nine species, including, in addition to *A. lipoferum* and *A. brasilense*, *Azospirillum amazonense* (Magalhaes et. al. 1983), *Azospirillum halopraeferens* (Reinhold et. al. 1987), *Azospirillum irakense* (Khammas et. al. 1989), *Azospirillum largimobile* (Ben Dekhil et. al. 1997), *Azospirillum doebereinerae* (Eckert et. al. 2001), *Azospirillum oryzae* (Xie and Yokota 2005) and *Azospirillum melinis* (Peng et. al. 2006).

Azosprillum has been found to colonize, promote growth, and increase the yield of numerous plant species (bashan and Levanony 1990; Bashan 1993; Okon and Labander-Gonzales 1994; Bashan and Holguin 1995). However, some of these effects can be enhanced when *Azospirillum* is co-inoculated with other microorganisms. A higher soybean yield was obtained while using mixed inoculants of *Azospirillum* and *Rhizobium* as compared to inoculations of *Rhizobium* alone (Singh and Subba Rao 1979). *Azospirillum*, by enhancing the proliferation of root hairs, increased the susceptibility of forage legumes to *Rhizobium* infectin (Yahalom et. al. 1987). A dual inoculation of A. *brasilense* and vesicular-arbuscular mycorrhizal fungi (VAM) increased the root biomass and the absorption of phosphorus by pearl millet (Subba Rao et. al. 1985), barley yield (Subba Rao et. al. 1985b) and the number of colonization sites of VAM fungi in halophytic plants growing in dunes (Will and Sylvia 1990).

N_2-fixation may be one of the minor mechanisms involved in plant-growth promotion by *Azospirillum* (Michiels et. al. 1989; Bashan and Levanony 1990). Despite the small importance of this mechanism to plants, the role of N_2-fixation in rhizocompetence, survival in soil and interactions of *Azospirillum* with other rhizosphere bacteria has been overlooked. The success of *Azospirillum* inoculants in promoting plant growth will largely depend on its survival in the hostile soil environment (Bashan et. al. 1995); and on its movement towards the host plant, both in bulk soil and in the rhizosphere (Bashan and Levanony 1987; Bashan and Holguin 1994). The capacity of a bacterium to fix N_2 may improve its survival as compared to non-fixing strains.

The interaction of N_2-fixing bacteria with other bacteria can inhibit or promote their diazotrophic activity, and this is quite common among microorganisms (Drozdowicz and Ferreira Santos 1987; Isopi et. al. 1995). The degradation of cellulose by *Cellulomonas* sp. CSI-17 provides *Azospirillum* sp. DN64 with a usable C source to obtain energy for N_2-fixation. The contribution of *Azospirillum* to *Cellulomonas* is fixed nitrogen (Halsall and Gibson 1989). The association between different *Azospirillum* species and the N_2-fixer *Bacillus polymyxa* enhanced the N_2-fixing activity of the co-cultures as compared to pure cultures of either *Azospirillum* or *Bacillus*. *Azospirillum* is benefited by the products which are released from the degradation of pectin by *Bacillus* (Khammas and Keiser 1992).

Holguin and Bashan (1993) reported that *Azospirillum* brasilense Cd fixed more N_2 when grown in a mixed culture with *Staphylococcus* sp., a non-N_2-fixing bacterium isolated from mangrove roots. This was not the result of either the increase in the bacterial population or of decreased O_2 concentration in the mixed culture. In mixed culture, the *Staphylococcus* population declined sharply, but not because *A. brasilense* Cd was more effective in competing for the available N. The addition of a cell-free dialyzate of *Staphylococcus* sp. culture medium to the *A. brasilense* culture significantly promoted the N_2-fixing capacity of the latter. When this dialyzate was produced by culturing Staphylococcus in N-free medium without yeast extract, the increased dialyzate activity depended on the concentration. When the dialyzate was diluted by volume to 50% and 25% of its original concentration, N_2-fixation by *A. brasilense* Cd increased significantly; when undiluted, the dialyzate failed to enhance N_2-fixation. Chemical analyses of the dialyzate by thin-layer chromatography identified aspartic acid; whereas, gas chromatography revealed succinic acid to be the major organic acid component. When artificially added to the *A. brasilense* Cd culture, only aspartic acid significantly promoted N_2-fixation by *A.* brasilense Cd. The N_2-fixing ability of *A. brasilense* Cd increased significantly when grown in mixed culture with the non-N_2 fixing bacterium *Staphylococcus epidermidis*, but not with *Micrococcus lylae*; both isolated from mangrove roots.

The N_2-fixing activity of *A. brasilense* Cd increases significantly when grown in mixed culture with the mangrove rhizosphere bacterium *Staphylococcus* sp. (Holguin and Bashan 1993).

Currently, the use of *Azospirillum* have been reported in crop management practices, due to N_2 fixation from the atmosphere by this bacterium, which improves the synthesis of auxin and gibberellins, stimulates root growth and thus, the absorption of water and nutrients, resulting in increases in productivity of several crops such as Oryza sativa, *Brachiaria* spp., *Saccharum officinarum*, *Zea mays* and *Triticum aestivum*, as well as reducing the amount of nitrogen fertilizer applied, or acts indirectly by protecting the plant pathogens present in soil, through the production of *siderophores, chitinases, glucanases* and *antibiosis*. Application of *Azospirillum* are reported in liquid form, peat, applying post-emergence as spraying, in seed furrows or use of pelleted seed, and the recommendation is the application directly in the seed coat in the form of liquid or peat on several crops (Ricardo et. al. 2013).

Azospirillum is a micro-aerobic associative type of nitrogen fixer present abundantly in the rhizosphere soil of C_4 type of grasses and plants. As early as 1925, Beijerinck, a Duch Microbiologist, observed the occurrence of *Spirillum lipoferum* (renamed as *Azospirillum lipoferum*) from soil. However, diazotrophic nature of this bacterium was realized and proved only in the year 1976 by Johna Dobereiner, a Brazilian lady scientist pioneering in associative dinitrogen fixation. This bacterium lives in close association with the roots of many cereal crops viz., maize, sorghum, pearl millet, finger millet, sugarcane, rice, foxtail millet and various types of grasses including *Cyanodon dactylon* – a weed. The species of *Azospirillum* so far reported are:

(1) *A. brasilense*

(2) *A. lipoferum*

(3) *A. amazonense*

(4) *A. seropedicae*

Comparing the efficiency of nitrogen fixation, *Azospirillum* is reported to fix nearly three times more nitrogen than *Azotobacter*. In addition to rhizosphere region, *Azospirillum* fixes nitrogen by invading upper cortical layer in the roots of C_4 type of plants in an associative symbiotic way. Most of the cereals, vegetables and all other non-leguminous crop plants are benefited from the inoculation with *Azotobacter* and/or *Azospirillum* biofertilizers. Crop plants inoculated have shown 15–20% increase in their yields as well as saving chemical N fertilizers 20–30%, in addition to increased dry matter and quality of food grains.

The important species used as biofertilizer are as follows:

Azospirillum lipoferum

In semisolid N-free medium, the cell size becomes wider (1.4–1.7μm) and longer (5–30 μm). Cells are ovide or pleomorphic and filled with phase refractile granules (probably Poly B-hydroxybutrate). The cells are gram negative, non-motile. Glucose and maltose used as sole carbon source for growth. Alkalinization of malate medium, due to oxidation of the malate, may be related to development of pleomorphism. Pleomorphism fails to occur when the organisms are cultured in semisolid N-free glucose medium which does not become alkaline. Bacteria posses a single flagellum when cultured on MPSS broth; however when cultured on MPSS agar at 30°C, numerous lateral flagella are formed in addition to the polar flagellum. The polar flagellum appears to be thicker than lateral flagella. Colonies of *Azospirillum* are pink, opaque, irregular, or round wrinkled and typically umbonate elevations. It possesses respiratory type of metabolism. Optimum temperature of 35–37°C is required for growth. The G + C content of DNA ranges from 69 to 70 moles %. Maize and other C_4 plants are selectively infected by this species, rather than by *A. brasilense*.

Azospirillum brasilense

In semisolid N-free medium, the bacterial cells are motile and vibrioid. Non-motile, enlarged, pleomorphic forms may also occur especially in older cultures or in association with the roots of grass seedling. A capsule is formed external to outer wall membrane of the cell. Colonies are deep pink, aerobic growth in liquid medium containing fixed nitrogen is usually homogenous, turbid and without clumping. No growth occurs aerobically in the absence of a source of fixed nitrogen. The G + C content of DNA ranges from 70 to 71 moles %. Wheat, barley, rice, oat and rye are selectively infected by nir strains of *A. brasilense*, rather than by *Azospirillum lipoferum*.

Azospirillum canadense

Azospirillum canadense (can.ad.en'se. N. L. neut. adj. *canadense* pertaining to Canada, the region of isolation, referring to its isolation from Canadian soil).

Cells are short rods, 0.9 × 1.8–2.5 µm in size, Gram-negative, motile via a single polar flagellum. White to light-pink, rounded, wet colonies form after 48–72 hours. Growth occurs on M medium at 20–37°C, pH 5–7 and 0.5–1.0% NaCl concentration. Optimum temperature is 25–30°C and optimum pH is 5–7. It is positive for nitrogen fixation and indole acetic acid production; negative for phosphate solubilization. Malic acid, potassium gluconate, acetic acid, pyruvic acid methyl ester, succinic acid monomethyl ester, cis-aconitic acid, citric acid, formic acid, D-galacturonic acid, D-glucuronic acid, α and β-hydroxybutyric acid, α-ketoglutaric acid, DL-lactic acid, malonic acid, propionic acid, quinic acid, D-saccharic acid, succinic acid, bromosuccinic acid, succinamic acid, D-alanine, L-asparagine and L-aspartic acid can be use as single carbon source.Sucrose, D-glucose, L-arabinose, D-arabitol, D-cellobiose, Lerythritol, D-fructose, L-fucose, D-galacose, gentibiose, myo-inositol, D-lactose, D-mannose, D-mannitol, maltose, D-melibiose, D-raffinose, L-rhamnose, D-sorbitol, D-trehalose, xylitol, D-gluconic acid, sorbitol, D-trehalose, xylitol, D-gluconic acid, α-ketobutyric acid, L-alanine, L-glutamic acid, L-histidine, L-leucine, L-ornithine, L-phenylalanine, L-proline, D-serine, L-serine, L-threonine, N-acetyl D-glucosamine, trisodium acetate, capric acid, adipic acid and phenylacetic acid are not utilized. Positive for catalase, oxidase, nitrate reduction, ß-glucosidase, ß-galactosidase and acetoin production and negative for indole production, arginine dihydrolase, urease and gelatin hydrolysis. Biotin is not required for growth. Major cellular fatty acids are 18:1 w7c, 16:1 w7c, 16:0. The DNA G + C content is 67.9 mol %. The predominant quinine system is ubiquinone Q-10 (Samina Mehnaz et. al. 2007). The type strain, DS2^T (=NCCB 100108^T=LMG 23617^T), was isolated from rhizosphere of corn (Zea mays) from Ontario, Canada.

Frankia

Much of the new nitrogen entering temperate forests come from bacteria, classified in the genus *Frankia*, that live in the root nodules of shrubs and trees. These root nodules, or "actinorhizal root nodules", are the major N$_2$-fixing symbioses in broad areas of the world. The symbiosis has become increasingly important, as climate changes threaten to remake the global landscape over the next several decades.

The genus *Frankia* was originally named by Jorgen Brunchorst in 1886 to honor the German biologist A. B. Frank. Brunchorst considered the organism that he had identified to be a filamentous fungus. Becking redefined the genus in 1970 as containing prokaryotic actinomycetes and created the family Frankiaceae within the Actinomycetales. He retained the original name of *Frankia* for the genus (http://web.uconn.edu).

Frankia sp. strains are filamentous bacteria that convert atmospheric N$_2$ gas into ammonia. This process is known as nitrogen fixation. *Frankia* fix nitrogen while living in root nodules of "actinorhizal plants". *Frankia* thus can supply most or all of the host plant's nitrogen needs. Consequently, actinorhizal plants colonize and often thrive in soils that are low in combined nitrogen.

Actinorhizal plants are a diverse group of woody species found on all continents (except Antarctica). Many are common plants, like alder, bayberry, sweet fern, etc., which one might pass every day. Others live in remote parts of the world. All play significant roles in the ecology of the soils in which they grow.

Actinorhizal plant includes the following:

- All species in the genus *Alnus* in the Betulaceae family.

- Some species in all four genera in the *Casuarinaceae* family.

- Certain species in the genus *Coriaria* in the Coriariaceae family.

- *Datisca cannabina* and *Datisca glomerata* in the Datiscaceae family.

- All species in the three genera in the Elaeagnaceae family. These are *Elaeagnus, Shepherdia* and *Hippophae*.

- All species in the genera *Myrica, Morella,* and *Comptonia* in the family Myricaceae.

- All species in six genera in the Rhamnaceae family. These are *Ceanothus, Colletia, Discaria, Kentrothamnus, Retanilla* and *Trevoa* and possibly *Adolphia*.

- Some species in the Rosaceae family, including all the species I the genera *Cercocarpus Cowania* and *Purshia* and some species of *Drays* (Shwintzer and Tjepkema 1990).

Frankia is a nitrogen-fixing gram positive bacterium that lives in the soil and has a symbiotic relationship with many plants "The actinomycete *Frankia* is of fundamental and ecological interests for several reasons including its wide distribution, its ability to fix nitrogen, differentiate into sporangium and vesicles (specialized cell for nitrogen-fixation) and to nodulatte plants from about 24 genera." (http://www.laspilitas.com/classes/Frankia.html).

The *actino*mycetous N_2-fixing endophytes *Frankia* are filamentous branching hypae. The diameter of hyphae varies between 0.5 and about 1 μm according to their age. *Frankia* are gram-positive organisms; but like other actinomycetes, this reaction may vary with the age of the culture. Callaham et. al. (1979) reported that the thinner hyphae of a *Frankia* strain isolated from *Comptonia peregrina* showed gram-negative staining, whereas the thicker hyphae were gram-positive.

In the *Frankia* genus, reproductive structures are called sporangia and generally develop at the end of the hyphae but may also do so in intercalary positions on hyphae filaments. They exhibit many different sizes and shapes. Mature club-shaped sporangia are 13–19 μm, but globular ones may reach 35 μm. Mature sporangia are divided into compartments that contain polyhedral sporangiospores 1.5–2.5 μm in diameter. The conditions governing the germination of sporangiospores are still unknown. *In vitro* and in most media, some *Frankia* strains readily produce specialized spherical structures, the vesicles, generally 2.5–3.0 μm in diameter. Other strains produce vesicles only in nitrogen-free media.

In addition to the three well-known structures mentioned above (hyphae, sporangia and vesicles), some *Frankia* strains isolated from *Casuarina* produce a fourth structure recently described by Diem and Dommergues (1984) for the first time. This new structure results from the conversion of a vegetative hypha into a wide torulose hypha with a thick double-layered wall. This type of hypha called a reproductive torulose hypha (RTH) may be disrupted into spore like cells and serves as the surviving and regenerating structure in *Frankia* strains from *Casuarina* when all vegetative hyphae are lysed. RTH cells are able to develop into new hyphae and give rise to the formation of colonies more rapidly than sporangiospores. RTH are thought to play major role in the reproduction, at least *in vitro,* of *Frankia* strains isolated from *Casuarina.*

Frankia sp. are filamentous nitrogen-fixing bacterium which grow by branching and tip extension, and thus resemble the antibiotic producing *Streptomyces* sp. They live in the soil and have a symbiotic relationship with certain woody angiosperms called actinorhizal plants. During growth, the *Frankia* sp. produce three cell types, i.e. sporangiospores, hyphae and diazo-vesicles (spherical, thick walled, lipid-enveloped cellular structures as stated above). The diazo-vesicles are responsible for the supplying of sufficient Nitrogen to the host plant during symbiosis. *Frankia* supplies most or all of the host plant nitrogen needs without added nitrogen and thus can establish a nitrogen-fixing symbiosis with host plants where nitrogen is the limiting factor in the growth of the host. Therefore, actinorhizal plants colonize and often prosper in soils which are low in combined nitrogen. Symbiosis of this kind adds a large proportion of new nitrogen to several ecosystems such as temperate forests, dry chaparral, sand dunes, mine wastes etc. (http://www. genoscope.cns.fr)

The *Frankia* alni ACN14a was first isolated in Tadoussac, Canada, from a green alder (*Alnus crispa*). Except for Australia and Antarctica, *Frankia alni* can be isolated from soils of all continents. *Frankia alni* causes root-hair deformation in a way that it enters the cortical cells and induces the nodule formations, which looks like those induced by *Rhizobium* in legumes. Then, the nodules are colonized by vegetative hyphae (mycelium filaments) that differentiate into diazo-vesicles. Reductive nitrogen fixation takes place in the diazo-vesicles, and this process is protected from molecular oxygen by many layers of tightly stacked hopanoid lipids.

The membranes of *Frankia*, as well as the membranes of some other bacteria like *Bradyrhizobium*, *Rhizobium* and *Streptomyces*, contain lipid components called hopanoids. Hopanoids, which are amphiphilic, pentacylic triterpenoid lipids, condense membrane lipids; therefore stabilizing the membranes in a similar way to which sterols do in higher organisms. There are many structural variants of hopanoids like polyol-and glycol derivatives that can be found within various bacteria; the specific functions of each hopanoid derivative have not been found. Few people hypothesize that hopanoids may have more refined functions than just membranes containing hopanoids surround vesicles called diazovesicles that contain nitrogenase, an oxygen-sensitive enzyme. While hopaniods thicken and stabilize the walls which help to keep oxygen away from the nitrogenase, few suggest that the hopanoids themselves play a more specific and molecular role in the oxygen protection mechanism of nitrogenase.

Frankia (both *Frankia* that lives in a symbiotic relationship with plants and free-living *Frankia* strains) secrete extracellular proteins which might be involved in processes like bacteriolysis, hydrolysis and virulence. Some *Frankia* strains have been known to secrete extracellular cellulases, pectinases and proteinases. It is not totally understood how these secretions impact *Frankia's* symbiotic systems, but they are being studied along with other common enzymes and proteins, like uptake hydrogenase, which are found in *Frankia's* symbiotic relationships. A hydrogenase is an enzyme that catalyses the reversible oxidation of molecular hydrogen (H_2). Hydrogen uptake is also present in free-living *Frankia*.

Organisms in symbiotic relationships with plant hosts are necessary for the health and survival of certain plants. They assist in creating and transporting certain root hormones, controlling pathogens and nematodes, root exploration, water retention, mineral uptake and resource sharing. *Frankia* specifically fixes nitrogen in the air and produces molecules that other plants can use.

Frankia has symbiotic relationships with numerous dicot plants and is said to be responsible for 15% of the biologically fixed nitrogen in the world. A type of symbiotic relationship including plant, mycorrhiza, and *Frankia* is called a tripartite relationship and is a complex and multi-layered community of organisms that protect and support each other. Other symbiotic relationships include *Frankia*, one or multiple plants and other bacteria. *Frankia alni's* specific role in these relationships is to infect the roots of plants; it deforms root hair of its plant host by going into the cortical cells and causing the formation of nodules. Vegetative hyphae colonize the nodules and then differentiate into diazovesicles. Diazovesicles are thick-walled (containing hopanoids) and spherical cells in which reductive nitrogen fixation occurs.

Frankia alni is the only named species in this genus, but there are a great many strains specific to different plant species. *Frankia alni* is a species of actinomycete filamentous bacterium that lives in symbiosis with actinorhizal plants in the genus *Alnus*. It is a nitrogen-fixing bacterium and forms nodules on the roots of alder trees. These are widely distributed in temperate regions of the northern hemisphere. The species of *Alnus glutinosa* is also found in Africa; and another, the Andean alder, *Alnus acuminate*, extends down the mountainous spine of Central and South America as far as Argentina. Evidence suggests that this alder may have been exploited by the Incas and used to increase soil fertility and stabilize terrace soils in their upland farming systems (Krajick 1998). *Alnus* species grow in a wide range of habitats which include glacial till, sand hills, the banks of water courses, bogs, dry volcanic lava flows, and ash alluvium (Schwencke and Caru 2001).

The first symptom of infection by *Frankia alni* is a branching and curling of the root hairs of the alder as the bacterium moves in. The bacterium becomes encapsulated with a material derived from the plant-cell wall and remains outside the host's cell membrane (Lalonde and Quispel 1977). The encapsulation membrane contains pectin, cellulose and hemicelluloses (Berg 1990). Cell division is stimulated in the hypodermis and cortex, which leads to the formation of a "pre-nodule". The bacterium then migrates into the cortex of the root while the nodule continues to develop in the same way as a lateral root. Nodule lobe primordial develops in the pericycle, endodermis, or cortex during the development of the pre-nodule and finally the bacterium enters the cells of these to infect the new nodule (http://web.uconn.edu). .

In culture and in some root nodules, multi-locular sporangia containing many spores are produced (Schwintzer and Tjepkema 1990). The sporangia are non-motile, but the spores can migrate to infect new host plants (http://web.uconn.edu). No air-borne dispersal of *Frankia alni* was detected, and it was thought that movement of water might account for the dispersal of the bacteria in peat soils (Arveby and Huss-Danell 1988).

In nitrogen-free culture and often in symbiosis, *Frankia alni* bacteria surround themselves in "vesicles". These are roughly spherical cellular structures that measure 2–6 mm in diameter and have a laminated lipid envelope. The vesicles serve to limit the diffusion of oxygen, thus assisting the reduction process that is catalyzed by the enzyme nitrogenase. This enzyme bonds each atom of nitrogen to three hydrogen atoms, forming ammonia (NH_3). The energy for the reaction is provided by the hydrolysis of Adenosine triphosphate (ATP). Two other enzymes are also involved in the process, glutamine synthetase and glutamate synthase. The final product of the reactions is glutamate, which is thus normally the most abundant free amino acid in the cell cytoplasm. A by-product of the process is gaseous hydrogen, one molecule of which is produced for every molecule of nitrogen reduced to ammonia, but the bacterium also contains the enzyme hydrogenase which serves to prevent some of this energy being wasted. In the process, ATP is recovered and oxygen molecules serve as the final electron acceptor in the reaction, leading to the lowering of ambient oxygen levels. This is to the benefit of the nitogenases, which only function anaerobically (http://web.uconn.edu).

As a result of their mutually beneficial relationship with *Frankia*, alder trees improve the fertility of the soils in which they grow and are considered to be a pioneer species, making the soil more fertile, and thus enabling other successive species to become established.

2.2.2 Phosphate solubilizing biofertilizers

Phosphorus (P) is the second major nutrient in crop productivity, as it plays a significant role in several physiological and biochemical activities such as cell division, photosynthesis, breakdown of sugar, transformation of sugar to starch, nutrient transport within the plant, transfer of genetic characteristics from one generation to another, and regulation of metabolic pathways (Tandon 1987; Armstrong 1988; Theodorou and Plaxton 1993). The maintenance of high level of soil phosphorus has been a major challenge to agricultural scientists, ecologists, and farm managers because in most of the soils, phosphate is present in unavailable form due to complex formation with Ca^{2+}, Al, Fe^{2+} or Mn^{2+} depending on soil pH and organic matter. The main problem of phosphorus in soil is its rapid fixation and the efficiency of P solubilization rarely exceeding 10–20%

The fixed forms of P in acidic soils are aluminium and iron phosphates, while in neutral to alkaline soils as calcium phosphates. The manufacture of phosphatic fertilizers requires high-grade rock phosphate (RP) and sulphur which are getting depleted progressively and becoming costlier. The total world reserves of RP are estimated to be around 2,700 billion tons of which 80% are located in the USA, Russia and Morocco. In India, RP deposits are estimated to be about 145 million tonnes, but bulk of this is of poor quality and is unsuitable for manufacture of phosphatic fertilizers. Only 25% of the total P requirement is met through indigenous sources; hence, about 1.5 million tons of high grade RP is imported annually.

The two types of phosphorous (P) in the soil are organic and inorganic. On an average, the amounts of P in the earth's crust and agricultural soils are 0.12 and 0.06%, respectively. Different parameters such as soil pH, calcium concentration, amount of organic matter, type and amount of clay, soil moisture, soil texture, root density and exudates can affect the availability of soil P to the plant (Tisdale et al. 1993; Barber 1995). Parameters including high pH, rich amount of $CaCO_3$, little amount of organic matter and drought decrease P availability to plants in the calcareous soils with arid and semiarid climates.

Using P fertilizers, especially superphosphate, as a very common method of providing plant P requirement, is not very efficient in calcareous and alkaline soils. Because under such conditions, high amounts of P are turned into insoluble products and become unavailable to the plant as only 20% of the fertilizer is soluble in the first year of use (Tisdale et. al. 1993). Application of P fertilizer has increased significantly to enhance crop yield production and as a result of using organic matter improperly. Use of rock phosphate as a source of P fertilizer, which is a simple and in the meanwhile not expensive method, is recommendable for acidic soils because in calcareous soils, high pH and high amount of $CaCO_3$ decreases the fertilizer solubility (Chein et. al. 1996; Abd-Elmonem and Amberger 2000). Different researchers have indicated that it is likely to increase P availability in soil. For example, acidizing rock phosphate, mixing rock phosphate with sulfur and organic matter and using rock phosphate with microorganisms including P solubilizing bacteria, sulfur oxidizing bacteria and arbuscular mycorrhiza are among the methods used for enhancing P availability (Chein 1996; Vessey 2003). Plant residues can be used as a source of C for soil fungi and heterotrophic bacteria, which produce organic acids enhancing P availability in the rock phosphate through protonization and chelating. Acid strength, the amount of soluble calcium and type and properties of chelating ligands are among the parameters affecting P availability (Chein et. al. 1996).

Many soil microorganisms especially *Pseudomona, Bacillus, Aspergillus* and *Penecillium* are effective in releasing P from inorganic and organic pool of total soil P through solubilization or mineralization and are known as phosphate solubilizing microorganisms. Phosphate solubilizing microorganisms (10% of total soil microorganisms), which include a large number of soil micro-flora (Whitelaw 1997; Sundara 2001), can solubilize inorganic phosphate (including soil phosphate) with the production of inorganic (carbonic and sulfuric) and organic (citric, butyric, oxalic, malonic, lactic) acids and phosphatase exzyme (Whitelaw 1997; Sundara et. al. 2001), which chelate the cations bound to phosphate via their hydroxyl and carboxyl group or by liberation of H^+, thereby converting it into soluble form. The activities of such microorganisms are affected by different soil parameters including soil fertility, temperature, moisture, organic matter and soil physical properties. Rock phosphate, Bacillus circulans and Cladosporium herbarum inoculants, mixture of rock phosphate and B. circulans and mixture of rock phosphate and C. herbarum resulted in the increased wheat dry weights. Schofield et. al. (1981) evaluated the use of 1:5 rock phosphate and elemental sulfur with Thiobacillus thiooxidance as a source of P fertilizer (biosuper) in three calcareous soils in the greenhouse. The enhanced P availability in rock phosphate combined with elemental sulfur has also been indicated by other researcher in which Thiobacillus sp. with elemental sulfur (biosuper) has been used (Stamford et. al. 2002). A large part of sulfur is biologically oxidized in the soil (Tabatabai 1986). Parameters affecting the P availability by rock phosphate when used in combination with elemental sulfur include the type of rock phosphate, the ratio of rock phosphate to elemental sulfur and crop and soil conditions (Rajan 2002). Elemental sulfur must be inoculated with Thiobacillus to enhance the P solubility of apatite. Soil P availability is determined by the following factors:

(1) The reaction time between apatite and organic acid

(2) The rate of organic acid dissociation

(3) The type and place of the functional (chemical) group

(4) The affinity of the chelating compound for cations

(5) Time and method of using rock phosphate

(6) Soil chemical and physical properties, especially the ability for P fixation

(7) Crop plant species and their nutritional requirements

(8) Soil particles size and their surface area

(9) Mineralogy and chemical properties of rock phosphate

(10) The activity and solubility of rock phosphate

(11) Soil organic matter (Grover 2003)

Apatite particles with size less than 0.15 are more beneficial to plant roots (Chein et. al. 2003). P solubilizing bacteria and rock phosphate have indicated their significant effects on crop yields such as wheat, rice and potato (Rajan 2002). PSM help in increasing the availability of accumulated phosphates in soil for plant growth by solubilization of phosphate, which in turn increases the biological nitrogen fixation efficiency on crop yield.

Phosphate solubilizing bioinoculants/biofertilizers are prepared from the bacteria or fungi which solubilize fixed form of phosphate in soil. Phosphatic bioinoculants are known by various names viz., Phosphobacterin, Microphos, Biophos, Bacteriophos, Phosphomicrobiales, etc. Phosphate solubilizing bioinoculants are prepared either as carrier-based inoculants from phosphate solubilizing bacteria, or as Sorghum grain-based inoculant from phosphate solubilizing fungi.

Phosphate solubilizing microorganisms include various species of bacteria, fungi, actinomycetes, and yeasts. The fungi are reported to be highly efficient phosphate solubilizers followed by bacteria and actinomycetes. Of the various microorganisms listed as phosphate solubilizers, the bacteria *Bacillus polymyxa* and *Pseudomonas striata* and fungi viz., *Penicillium digitatum* and *Aspergillus awamori* are used in commercial preparations. Compared to bacteria, fungal species are known to solubilize more phosphate. Generally, soil pH of 6.5–7.0 and 7.0–7.5 is favorable for phosphate solubilization by fungal and bacterial isolates, respectively.

Phosphate solubilizing microbes

Bacteria: *Bacillus polymyxa*, *B. megetherium* var *phospheticum*, *B. megetherium* var. *serratia*, *B. circulens*, *Pseudomonas striata*, *P. liquifaciens*, *Achromobacter* spp., *Arthobacter* spp.

Fungi: *Penicillium digitatum*, *Aspergillus awamori*, *A fumigatus*, *Penicillium digitatum*, *P. liliacinum*, *Cephalosporium* spp., *Trichoderma* spp.

Actinomycetes: *Streptomyces* spp., *Nocardia* spp.

Yeast: *Rhodotorula* spp., *Schwanniomuces occidentails*

Phosphate solubilizing biofertilizers are prepared from the microorganisms which solubilize fixed phosphorus in soil and make it available to the plants. The most of P in the form of phosphatic fertilizers added to soil becomes unavailable to the plants due to phosphate fixing capacity of soil. To benefit fully from added P fertilizers in soil, new technology has come forward in the form of use of phosphate solubilizing biofertilizers. Phosphate solubilizing bacteria can function well in the neutral to alkaline soils and the fungi can function well in acidic soils. Extensive field trials on use of carrier-based preparations of phosphate solubilizing biofertilizers indicated the beneficial effects on P uptake, growth and yield of a vast range of legumes, cereals and vegetable crops (Wani 1980). Phosphatic biofertilizers have been reported more effective when used in combination with N-fixing biofertilizers. Lignite based preparations of phosphatic biofertilizers are recommended for use as seed treatment, while sorghum-grain based phosphatic biofertilizers are used for addition in compost pit or added along with compost for horticultural crops.

Bacillus polymyxa

Cells are rod shape with 0.6–0.8 to 2.0–5.0 μm in size. These are motile with typically lateral flagella and gram-negative. Optimum temperature of 30°C and pH 6–7 is required for growth. Spore has parallel, longitudinal surface ridges so that it is star-like in cross-section. Colonies on nutrient agar are thin, often with amoeboid spreading; whereas on glucose agar, these are usually heaped and mucoid with matt surface. Growth is generally adherent to agar medium. Nitrogen is fixed under anaerobic condition. Lactose is utilize as carbon source, where as NH_4 is used as nitrogen source. Anaerobic growth is vigorous in presence of fermentable carbohydrate. The G + C content of DNA ranges from 43 to 46 moles %.

2.2.3 Vesicular arbascular mycorrhizae

Tulasne and Tulasne (1841) was the first to locate the tree roots surrounded by truffle fungi. Vittadini (1842) found their beneficial role with plant roots and proposed the hypothesis that tree rootlets are nourished by certain fungal mycelia. Frank further in 1885 referred this hypothesis to a theory of mutualistic symbiosis. He coined the term "mycorrhiza" for the fungus root. The mycorrhizal fungi form a symbiotic relationship with plant roots. The fungus takes carbohydrates from the plant and in turn supplies the plant with the nutrients, growth promoting harmones, etc., and protects the host partner from the root pathogens.

VA-mycorrhizae are the non-septate mycelia phycomycetous fungi. VA-mycorrihal genera are *Acaulospora*, *Entrophosphora*, *Gigaspora*, *Glomus*, *Sclerocystis* and *Scutellspora*. The common species are *Acaulospora laevis*, *A. rugosa*, *A. trappie*; *Entrphosphora colombiana*, *E. infrequens* and *E. schenckii*; *Gigaspora margarita*, *G. gigantean* and *G. albida*; *Glomus mosseae*, *G. fasciculatum*, *G. intradices*, *G. pallidum*, *and G. monosporum*, *Sclerocystis clavispora*, *S. sinuosa*, *S. microcarpus*, *Scutellspora nigra and S. alborosea*. Amongst these, *Glomus* species are found worldwide.

In addition to phosphate solubilizers, plants are also benefited by mycorrhizae which help them in improving P nutrition. Amongst the various mycorrhizae, vesicular arbuscular mycorrhizae mobilize soil phosphate by remaining in mutual association with plant root system and enhance the P uptake of host plant. *Gigaspora calospora*, *Glomus fasciculatum and Gigsspors margarita* are the important species of VA-mycorrhizae.

Amongst the various endomycorrhizae, vesicular arbuscular mycorrhizae (VAM) are found to occur within the roots of most of the food, vegetables, and horticultural crops and tropical trees. They colonize the roots of more plants than any other mycorrhizae and are necessary for many plants to grow well. The most common mycorrhizal association is the vesicular arbuscular mycorrhizae which produce the structures such as vesicles and arbuscles in root cortex region of the crop plants. They are obligate symbionts and could not be cultured on nutrient media. They produce extrametrical mycelia which travel many centimeters away from the root and absorb nutrients and translocate them to the host cells.

An extensive research work on utilization of VAM technology has been carried out all over the world. Tremendous pot and field experiments conducted on VAM revealed that they improved growth, nutrient uptake and yield of cereals, pulses, vegetables (except cruciferous crops), plantation crops, ornamental and forest trees. The crops which are from *Allium* viz., *Allium sativum* and *A. cepa* were found to be highly responsive to mycorrhizal inoculants. Transplanted vegetable crops like tomato, chilli, onion, brinjal, etc., find wider scope and feasibility for utilization of VA-mycorrhizal inoculants. The VAM inoculum requirements for such transplanted crops is minimized by growing the seedlings with mycorrhizal inoculums and such mycorrhizal seedlings when transplanted have been found to show remarkable increase in plant growth, nutrient uptake and yields.

VA-mycorrhizae enhance P uptake of crop plants and help in better utilization of applied phosphatic fertilizers, particularly in the soils subjected to rapid fixation of P. The extrametrical mycelium produced by these fungi reaches several centimeters away from the rhizosphere zone, absorbs nutrients from beyond depletion zone and supplies the host cells with phosphorus. VA-mycorrhizal colonization increases plant growth also by enhancing the uptake of various micronutrients viz., Zn, Fe, Cu, Mn, Ca, S, etc. VAM association helps the host crop plants in controlling the soil borne plant pathogens. VA-mycorrhizae induce the changes in root exudation, especially increase of the arginine content and in thickening of cortical root-cell walls which account for the resistance of plant roots against soil borne plant pathogenic fungi and plant parasitic nematodes. VA-mycorrhizae absorb moisture far away from root surface and also from low level efficiently and allow survival of crop plants under drought conditions. The synergistic relationship of VAM has been reported with other useful soil microorganisms like *Azotobacter*, *Azospirillum*, *Rhizobium* and P solubilizing bacteria and fungi. The dual inoculations of VAM with either of these microbes have been reported more beneficial than their single inoculations.

Mycorrhizal association is considered crucial for the survival and the growth of the majority of plant species in natural ecosystem (Harley and smith 1983). The role of mycorrhiza in enhancing water and nutrient uptake, especially phosphorous, zink and copper are well known (Bowen et. al. 1974). In many tropical soils, lack of phosphate is most important component affecting plant growth especially when the soils are degraded acid soils like in Kerala. In such soils, the success of forestation programmed depends upon the establishment of seedlings. High colonization of roots by VAM Fungi is species specific and is mycorrhiza dependent. Interestingly for each tree species, the efficient VAM species are different. Increase in growth, shoots and root dry weight and P content of inoculated plants shows the positive role of VAM in promoting growth which will help in the successful establishment of plantations in degraded soil.

Mycorrihizae have the potential to contribute significantly to the success or failure of agro-ecosystems. For example, Reeves et. al. (1979) reported that in a semi-arid environment, more than 99% of the plant cover in a natural community was VA mycorrhizal; while in a disturbed soil, less than 1% of the plant cover was mycorrhizal. Allen (1989) described an apparently pathogenic effect of mycorrhizal fungi on non-mycotrophic species, which could help to explain the abundance of non-mycotrophic plants in severely disturbed soils where inoculums density is low as well as their relative scarcity in soils of high inoculums. Allelopathic inhibition of VAM host plants by ponderosa pine has been suggested as a possible explanation for the lack of VAM inoculums in sites with pines (Kovalcic et. al. 1984).

The species of mycorrhizal fungus present in the soil can influence the competitive abilities of plant species. In the absence of mycorrhizal fungi, only the species, which do not need mycorrhizae, will be able to grow. Therefore, non-mycorrhizal species are most likely to dominate plant communities on poor soils containing no or few mycorrhizal fungi propagules (Bagyaraj 1989). Bethlenfalvay et. al. (1989) suggest that geographically isolated populations of VAM (fungal species) develop distinct characteristics capable of eliciting different symbiotic response and introduce the concept of edophytes to designate such intraspecific variations of these soil fungi. The fungi of VAM are present in almost all undisturbed soils, but may be lost following mechanical disturbance and removal of the vegetation (St. John 1990). Drought and soil erosion are other factors which may reduce or eliminate the VAM fungi propagules (Powell 1980).

Recently, a reduction in VAM inoculums potential due to mound formations by golphers was reported (Koide and Mooney 1987). Therefore, topsoil disturbance decreases the number of spores or the number of spore types that could be isolated from the soil and reduces or delays formation of VA mycorrhizae (Jasper et. al. 1987). Moorman and Reeves (1979) reported only 2% infection in a disturbed soil as compared to 77% in an adjacent undisturbed soil.

It is easier to list out the plant families which do not form VA-mycorrhizal association than to list those that do. VAM association is reported in the majority of agricultural crops, most shrubs, most tropical and few temperate tree species. Plant families which do not form VA-mycorrhizae are found in most angiosperms, some gymnosperms and in pteridophytes and briophytes. There are also mycorrhizae of the economically important crops, i.e. legumes, maize, wheat, barley, rice and other cereals, vegetable crops, temperate fruit trees, many tropical timber trees, woody shrubs, tropical plantation crops (cocoa, coffee, tea, rubber, etc.), cotton, tobacco, olive, citrus and grapevine. In addition, the VA-mycorrhizal association with plants is geographically ubiquitous and occurs in plants growing in arctic, temperate, and tropical regions in all types of soils. VA-mycorrhizae occur over a broad ecological range from aquatic to desert environment.

Despite the non-specificity of VAM fungi with respect to host plant, certain fungus-plant combinations are more efficient than others. Dela Cruz et. al. (1988) have reported that *Sclerocystys clavispora* was consistently ineffective in promoting growth of the three hosts: *Acacia auriculiformis*, *A. mangium* and *Albizia falcataria*. Nodulation and mycorrhizal infection in plants inoculated with this fungus were poor and similar to that of uninoculated plants. However, noducation and N_2 fixation were high in all plants inoculated with the VAM fungi *Glomus fasciculatum*, *G. margarita* and *Scutelospora persica*.

2.2.4 Potassium solubilizing biofertilizer

Potassium is one of the essential macronutrient and the most abundantly absorbed cation in higher plants. It plays an important role in the growth and development of plants. It activates enzymes, maintains cell turgor, enhances photosynthesis, reduces respiration, helps in transportation of sugars and starches, helps in nitrogen uptake and is essential for protein synthesis. In addition to plant metabolism, potassium improves crop quality because it helps in grain filling and kernel weight, strengthens straw, increases disease resistance and helps the plant better to withstand stress. The introduction of high yielding varieties and hybrids during green revolution and with the progressive intensification of agriculture, the soils are getting depleted in potassium reserve at a faster rate. As a consequence, potassium deficiency is becoming one of the major constraints in crop production.

There are three forms of potassium found in the soil viz., soil minerals, non-exchangeable and available form. Soil minerals make up more than 90–98% of soil potassium. It is tightly bound and most of it is unavailable for plant uptake. The second is non-exchangeable potassium which acts as a reserve to replenish potassium taken up or lost from the soil solution. It makes up approximately 1–10% of soil potassium. The third type is available potassium which constitutes 1–2%. It is found either in the solution or as part of the exchangeable cation on clay mineral. Among three different forms of potassium in soils, the concentrations of soluble K in soils are usually very low; but the highest proportion of potassium in soils is in insoluble rocks and minerals (Goldstein 1994). A significant share of soil potassium occurs in unavailable form in soil minerals such as orthoclase and microcline (K-feldspars).

Although K deficiency is not as wide spread as that of nitrogen and phosphorus, many soils which were initially rich in K became deficit in due course due to heavy utilization by crops and inadequate K application, runoff, leaching and soil erosion (Sheng and Huang 2002). Potassium deficiency symptoms usually occur first on the lower leaves of the plant and progress towards the top as the severity of the deficiency increases. One of the most common signs of potassium deficiency is the yellow scorching or firing (chlorosis) along the leaf margin. In severe cases of potassium deficiency, the fired margin of the leaf may fall out. Potassium deficient crops grow slowly, root systems are poorly developed, stalks are weak, and there is lodging of cereal crops.

Microorganisms play a key role in the natural K cycle. Few species of rhizobacteria are capable of mobilizing potassium in accessible form in soils. There is considerable population of K solubilizing bacteria in soil and rhizosphere (Sperberg 1958). Silicate bacteria were found to dissolve potassium, silicon, and aluminium from insoluble minerals (Aleksandrov et. al. 1967). It has been reported that most of potassium in soil exists in the form of silicate minerals. The potassium is made available to plants when the minerals are slowly weathered or solubilized (Bertsch et. al. 1985). Potassium solubilizing bacteria are capable of solubilizing rock K and mineral powder such as mica illite and orthoclases through production and excretion of organic acids (Friedrich et. al. 1991).

The microorganisms like bacteria, fungi and actionmycetes were colonized even on the surface of mountain rocks (Gromov 1957). Norkina and Pumpyanskaya (1956) reported that the silicate solubilizing bacteria B. mucilaginosus subsp. siliceus liberates potassium from feldspar and aluminosilicates.

Aleksandrov et. al. (1967) isolated different bacterial species like silicate bacteria which were found to dissolve potassium, silica and aluminium from insoluble minerals. Heinen (1960) reported the ability of *Bacillus caldolyticus* and *Proteus* sp. to grow and soublize quartz. Avakyan et. al. (1986) and Li (1994) isolated K solubilizing bacteria from soil, rock, and mineral samples and identified it as B. mucilaginosus based on morphological and physiological characters. Potassium solubilizing rhizobacteria were isolated from the roots of cereal crop by the use of specific potassium bearing minerals (Mikhailouskaya and Tchernysh 2005). Hu et. al. (2006) reported K solubilizing strains from the soil which effectively dissolve mineral potassium when they are grown on Alekandrove medium. Sugumaran and Janarthanam (2007) isolated K solubilizing bacteria from soil, rocks and minerals samples viz., orthoclase, Muscovitemica. Among the isolates, B. mucilaginosus solubilize more potassium by producing slime in muscovite mica.

Sheng and Huang (2002) observed potassium release from strains of potassium soublizing bacteria at 35.2 mg/l in 7 days at 28°C at pH range from 6.5–8.0. Badr (2006) studied extent of potassium and phosphorus solubilization by silicate solubilizing bacteria. It ranged from 490 mg/l to 758 mg/l at pH 6.5–8.0 and the potassium solubilization by B. mucilaginosus isolated from soil, rock, and mineral samples recorded 4.29 mg/l release of potassium in media supplemented with muscovite mica.

The optimum growth condition for potassium solubilizing bacteria is at temperature of 32°C and at pH 8. In acidic soils, soil fertility and crop productivity decreases because of limited availability of essential nutrients, especially potassium and accumulation of toxic elements such as Aluminium (AI) and Magnesium (Mn). Soil microbes play an important role in maintaining soil fertility and productivity.

Mechanisms of Potassium solubilization is reported by various workers. Moira et. al. (1963) isolated many of the fungal isolates having the potential to release metal ions and silicate ions from minerals (saponite and vermiculite), rocks and soils. The fungal isolates are known to produce citric acid, oxalic acid decomposed or solubilized natural silicates, and help in removal of metal ions from the rocks and soils.

Berthelin (1983) reported potassium solubilization from precipitated forms through production of inorganic and organic acids by Thiobacillus, Clostridium and Bacillus. The mineral dissolution was enhanced due to production of mucilaginous capsules containing exopolysaccharides. Freidrich et. al. (1991) and Ullman et. al. (1996) reported that potassium solubilizing bacteria B. mucilaginosus were able to solublize rock K mineral powder such as micas, illite and orthoclases through production and excretion of organic acids. Styriakova et. al. (2003) reported the activity of siliate dissolving bacteria to play a pronounced role in the release of Si, Fe and K from feldspar and Fe hydroxides. Sheng and Le (2006) also reported that solubiliztion of illite and feldspar by microorganisms was due to the production of organic acids like oxalic acid and tartaric acids and also due to production of capsular polysaccharides which help in dissolution of minerals to release potassium.

K-solubilization ability can be determined by Aleksandrov agar medium containing potassium aluminosilicates as insoluble source of K-minerals.

Potassium solubilizing bacteria produces IAA or their derivatives, and 43.5% of the strains displayed siderophores. Some strains showed better growth on feldspar and effects on solubilization. Two strains of *Bacillus mucilaginosus*, i.e. AFM2 and AC2 exhibited greater K releasing efficiency than other isolates.

Various workers have reported the effect of inoculation of potash solubilizers on plant growth and yield of several crop plants. Khudsen et. al. (1982) and Krieg and Holt (1984) isolated potassium solubilizing bacteria from rock and mineral samples which showed higher activity in potassium release from acid leached soil and improved green-gram seedling growth. Datta et. al. (1982) and Nianikoval et. al. (2002) observed that Phosphorus solubilizing bacteria and silicate bacteria play an important role in plant nutrition through the increase in P and K uptake by plant. Xue et. al. (2000) and Sheng et. al. (2003) reported silicate dissolving bacteria could improve soil P, K, Si reserves and promote plant growth. Nayak (2001) reported the effect of potash mobilizer on brinjal and recorded an increased potash uptake and increased plant biomass in potash mobilizer treated plants as compared to the control plants. Lin et. al. (2002) recorded increase in biomass by 125%, K and P uptake more than 150% in tomato plant due to inoculation of silicate dissolving bacteria (B. *mucilaginosus*) than the non-inoculation, and opinioned that there is a potential in applying this bacteria's strain RCBC 13 for improving K and P nutrition. Vessey (2003) reported the use of plant growth promoting rhizobacateria (PGPR) including phosphate and potassium solubilizing bacteria (PSB and KSB) as biofertililzer, as a sustainable solution to improve plant nutrient and production.

Park et. al. (2003) reported that bacterial inoculation could improve phosphorus and potassium availability in the soils by producing organic acid and other chemicals and stimulating growth and mineral uptake of plants.

Sheng et. al. (2003) studied the effect of inoculation of sulphur solubilizing bacteria (Bacillus endaphicus) on chilli and cotton which resulted in increase in available P and K contents in plant biomass. Styriakova et. al. (2004) reported that *Acidothiobacillus ferroxidans* cultures enhanced the chemical dissolution of the mineral and formed partially weathered interlayer from where K was expelled. This was coupled with the precipitation of K and Jarosite. Zhang et. al. (2004) reported the effect of potassic bacteria on sorghum, which results in increased biomass and contents of P and K in plants. Clarson (2004) reported the increased K uptake coupled with increased yield in yam and tapioca in the plants with potassium mobilizer in conjunction with chemical fertilizers. Chandra et. al. (2005) reported increased yield by 15–20% in yam and tapioca due to the potash solubilizer application in combination with other biofertilizers like *Rhizobium, Azospirillum, Azotobacter, Acetobacter* and PSM. Wu et. al. (2005) found inoculation of K solubilizer (*B. mucilaginosus*) along with P solubilizer (*B. megaterium*) and N-fixer (*Azotobacter chroococcum*) increased the growth, nutrient uptake significantly in maize crop and also improved soil properties such as organic matter content and total N in soil. Han and Lee (2005) found that the co-inoculation of PSB and KSB in combination with direct application of rock P and K materials into the soil resulted in increased P and K uptake, photosynthesis and the yield of egg plant grown on P and K limited soil. Ramarethinam and Chandra (2005) in a field experiment recorded increased brinjal yield, plant height and K uptake significantly compared to control due to inoculation of potash solubilizing bacteria (*Frateuria aurantia*). Mikhailouskaya and Tchernysh (2005) reported the effect of inoculation of K mobilizing bacteria on severally eroded soils which are comparable with yields on moderately eroded soil without bacterial inoculation, in increased wheat yield up to 1.04 t/ha. Sheng (2005) worked on potassium releasing bacteria B. edaphicus for plant growth promoting effects and nutrient uptake on cotton and rape seed in K deficient soil. Pot experiments resulted in increased root and shoot growth and potassium content was increased by 30–26%, respectively. In chilli crop, increased biomass and K uptake was due to inoculation of potash solubilizer (Ramarethinam and Chandra 2005). Christophe et. al. (2006) reported that *Burkhulderia glathei* in association with pine roots significantly increased weathering of biotite and concluded that there was effect of *B. glathei* PMB and PML on pine growth root morphology which was attributed to release of K from the mineral.

Sheng and Le (2006) recorded an increased root and shoot growth with significantly higher N, P, and K contents of wheat plant components due to inoculation of B. edaphicus in a yellow-brown soil which had low available K. In the field experiment, Badr (2006) recorded increased yield in tomato crop due to inoculation of silicate dissolving bacateria B. cereus as bioinoculant along with feldspar and rice straw on K releasing capacity. Han et. al. (2006) evaluated the potential of PSB and KSB inoculated in nutrient limited soil planted with pepper and cucumber. The co-inoculation of PSB and KSB showed high P and K content and plant growth compared to control.

Supanjani et. al. (2006) reported that integration of P and K rocks with inoculation of phosphorus and potassium solubilizing bacteria increased P availability from 12 to 21% and K availability from 13 to 15% in the soil as compare with control and subsequently improved nutrient N, P and K uptake in *Capsicum annuum*. The integration also increased plant photosynthesis by 16% and leaf area by 35% as compared to control. On the other hand, the biomass harvest and fruit yield of the treated plants were increased by 23–30%, respectively. Overall result of this finding is the treatment with P and K rocks; and P and K solubilizing bacterial strain was sustainable alternative to the use of chemical fertilizer.

The potential phosphate solubilizing bacteria (PSB) B. megaterium var, *Phosphaticum* and potassium solubilizing bacteria (KSB), *B. mucilaginosus* were evaluated using pepper and cucumber as test crops. The rock phosphorus and potassium applied either singly or in combination do not significantly enhance availability of soil phosphorus and potassium indicating their unsuitability for direct application. Co-incubation of PSB and KSB with the rock phosphorus and potassium resulted in consistently higher P and K available than in the control (Vassilev et. al. 2006). Sugumaran and Janarthanam (2007) recorded increase in the dry matter by 125% and oil content by 35.4% in groundnut plant and available P and K was increased from 6.24 to 9.28 mg/kg and 86.57 to 99.60 mg/kg respectively in soil due to inoculation of B. mucilaginosus (KSB) compared to uninoculated control.

2.2.5 Sulfur oxidizing biofertilizer

Sulfur (S) is an essential element for many plant functions. It is a structural component of protein and peptides and various enzymes. Activate conversion of inorganic N into protein promotes nodule formation in legumes, a catalyst in chlorophyll production and a structural component of the compounds that gives the characteristic odor and flavors to mustard, onion and garlic. Since these are building blocks in the plants, sulfur becomes an important component in the plant structure. Therefore, their deficiency in plant shows the deficiency symptoms. The classical symptom of deficiency is a paleness/whitishness of younger foliage. The use of concentrated NPK fertilizers without S in its composition, coupled with the largest exportation of nutrients by crops and reduction of atmospheric S inputs is resulting in increased agricultural areas deficient in S in the world.

One possibility for the replacement of sulfur in the soil is the use of elemental sulfur (SO). However, before being absorbed by plants, SO must be microbiologically oxidized to sulfate. The oxidation of SO was always associated with the species Acidithiobacillus thiooxidans and *A. ferrooxidans*; although a great diversity of organisms have this capacity in the soil. The SO oxidation rate in soil is influenced by abiotic factors, but the main factor of variation is the diversity of native microbial community.

Biological oxidation of hydrogen sulfide to sulfate is one of the major reactions of the global sulfur cycle. Reduced inorganic sulfur compounds are exclusively oxidized by prokaryotes and sulfate is the major oxidation product.

The sulfur-oxidizing prokaryotes are phylogenetically diverse. In the domain, *Archaea* aerobic sulfur oxidation is restricted to members of the order *Sulfolobales* (Fuchs et. al. 1996; Stetter et. al. 1990) and in the domain *Bacteria* sulfur is oxidized by aerobic lithotrophs or by anaerobic phototrophs. The non-phototrophic obligate anaerobe *Wolinella succinogenes* oxidizes hydrogen sulfide to polysulfide during fumarate respiration (Macy et. al. 1986). The ecology, physiology and biochemistry of sulfur-oxidizing neutrophilic chemolithotrophic bacteria have been reviewed (Kelly 1982; Kelly 1989; Kelly et. al. 1997); the acidophilic sulfur-oxidizing bacteria have been reviewed by Harrison (1984) and Pronk et. al. (1990); and the molecular genetics of *Acidithiobacillus ferrooxidans* has been reviewed by Rawlings and Kusano (1994). The sulfur metabolism of phototrophic bacteria has been reviewed by Brune (1989; 1995) and Truper and Fischer (1982). The physiology and genetics of both phototrophic and lithotrophic sulfur-oxidizing prokaryotes have been reviewed by Friedrich (1998).

Prokaryotes oxidize hydrogen sulfide, sulfur, sulfite, thiosulfate and various polythionates under alkaline (Sorokin et. al. 2001), neutral, or acidic conditions (Harrison, 1984). Aerobic sulfur-oxidizing prokaryotes belong to genera like *Acidianus* (Friedrich 1998), *Acidithiobacillus* (Kelly 2000), *Aquaspirillum* (Friedrich 1998), *Aquifex* (Huber and Stetter 1999), *Bacillus* (Aragno 1991), *Beggiatoa* (Strohl and Genus), *Methylobacterium* (de Zwart et. al. 1996; Kelly and Smith 1990), *Paracoccus*, *Pseudomonas* (Friedrich and Mitrenga 1981), *Starkeya* (Kelly et. al. 2000), *Sulfolobus* (Stetter et. al. 1990), *Thermithiobacillus* (Kelly and Wood 2000), *Thiobacillus* and

Xanthobacter (Friedrich and Mitrenga 1981) and are mainly mesophilic. Phototrophic anaerobic sulfur-oxidizing bacteria are mainly neutrophillic and mesophilic (Brune 1995; Suylen et. al. 1986) and belong to genera like *Allochromatium* [formerly *Chromatium* (Imhoff et. al. 1998)], *Chlorobium, Rhodobacter, Rhodopseudomonas, Rhodovulum and Thiocapsa* (Brune 1989). Lithoautotrophic growth in the dark has been described for *Thiocapsa roseopersicina, allochromatium vinosum* and other purple sulfur bacteria, as well as for purple non-sulfur bacteria like *Rhodovulum sulfidophilum* (formerly *Rhodobacter sulfidophilus*) (Hiraishi and Umeda 1994), *Rhodocyclus genatinosus* and *Rhodopseudomnas acidophila* (Kondratieva 1989; Siefert and Pfennig 1979).

Sulfur oxidation involves the oxidation of reduced sulfur compounds such as sulfide (H_2S), inorganic sulfur (S_0) and thiosulfate ($S_2O_2^{-3}$) to form sulfuric acid (H_2SO_4). An example of a sulfur-oxidizing bacterium is Paracoccus.

Generally, the oxidation of sulfide occurs in stages, with inorganic sulfur being stored either inside or outside of the cell until needed. The two step process occurs because sulfide is a better electron donor than inorganic sulfur or thiosulfate; this allows a greater number of protons to be translocated across the membrane; sulfur-oxidizing organisms generate reducing process that pushes the electrons against their thermodynamic gradient to produce NADH. Biochemically, reduced sulfur compounds are converted to sulfite (SO_2^{-3}) and, subsequently, sulfate (SO_2^{-4}) by the enzyme sulfite oxidase. Some organisms, however, accomplish the same oxidation using a reversal of the APS reductase system used by sulfate-reducing bacteria. In all cases, the energy liberated is transferred to the electron transport chain for ATP and NADH production. In addition to aerobic sulfur oxidation, some organisms (e.g., Thiobacillus denitrificans) use nitrate (NO^{-3}) as a terminal electron acceptor and therefore grow anaerobically.

A classic example of a sulfur-oxidizing bacterium is beggiatoa, a microbe originally described by Sergei Winogradsky. Beggiatoa can be found in marine or freshwater environments. They can usually be found in habitats that have high levels of hydrogen sulfide. These environments include cold seeps, sulfur springs, sewage contaminated water, mud layers of lakes, and near deep hydrothermal vents. Beggiatoa can also be found in the rhizosphere of swamp plants. Winogradsky found that Beggiatoa oxidized hydrogen sulfide (H_2S) as an energy source, forming intracellular sulfur droplets. Winogradsky referred to this form of metabolism as inorgooxidation (oxidation of inorganic compounds). The finding represented the first discovery of lithotrophy.

Beggiatoa can grow chemoorgano heterotrophically by oxidizing organic compounds to carbon dioxide in the presence of oxygen; though, high concentrations of oxygen can be a limiting factor. Organic compounds are also the carbon source for biosynthesis. Few species may oxidize hydrogen sulfide to elemental sulfur as a supplemental source of energy (facultatively litho-heterotroph). This sulfur is stored intracellularly. Few species have the ability of chemolithoautotrophic growth, using sulfide to elemental sulfur as a supplemental source of energy (facultatively litho-heterotroph). This sulfur is stored intracellularly. Few species have the ability of chemolithoautotrophic growth, using sulfide oxidation for energy and carbon dioxide as a source of carbon for biosynthesis. In this metabolic process, internal stored nitrate is the electron acceptor and it is reduced to ammonia.

Sulfide oxidation: $2H_2S + O_2 \rightarrow 2S + 2H_2O$.

Marine autotrophic Beggiatoa species are able to oxidize intracellular sulfate. The reduction of elemental sulfur frequently occurs when oxygen is lacking. Sulfur is reduced to sulfide at the cost of stored carbon or by added hydrogen gas. This may be a survival strategy to bridge periods without oxygen.

The bacteria belonging to the families of Thiobacteriaceae, Beggiatoaceae and Achromatiaceae, known as colorless sulfur bacteria, has the ability to oxidize reduced inorganic sulfur. The oxidation of inorganic sulfur compounds is carried out by a spectrum of sulfur-oxidizing organisms which includes the following:

(1) Obligate chemolithotrophic organisms

(2) Mixotrophs

(3) Chemolithotrophic heterotrophs

(4) Heterotrophs which do not gain energy form the oxidation of sulfur compounds, but benefit in other ways form this reaction

(5) Heterotrophs which do not benefit from the oxidation of sulfur compounds

The spectrum is completed by a hypothetical group of heterotrophic organisms, which may have a symbiotic relationship with thiobacilli and related bacteria. Such heterotrophs may stimulate the growth of colorless sulfur bacteria and thereby contribute to the oxidation of sulfur compounds. The liberation of S from organic compounds provides sulfite. This provides an increase in soil acidity, but most pronounced accumulation of acid by microorganisms is by the oxidation of sulfides that have accumulated in soils during water logging and anaerobic conditions. Several models of the atmospheric sulfur cycle have been proposed (Eriksson 1963; Robinson and Robbins 1968, 1970; Kellogg et. al. 1972; Friend 1973; Granat et. al. 1976). All indicate that a large proportion (usually about 50%) of the sulfur in the atmosphere is derived from biological transformations of sulfur in the pedosphere and hydrosphere and that most of the sulfur volatilized from natural systems through microbial activity is in the form of H_2S.

Some carrier based and liquid based S biofertilizer bioformulation are available. These are Thiobacillus biofertilizer, symbion-S, and micro S-109.

Thiobacillus biofertilizer

Thiobacillus biofertilizer consist of the most effective sulfur oxidizing bacteria which oxidizes elemental sulfur in the shortest time and in large quantities. Oxidation of sulfur not only lowers soil pH creating an environment where plants can easily absorb required nutrients, but also produces SO_4 to be taken up by plants, a nutrient which is of so great importance to the health and nutrition of crops.

Symbion-S

Symbion-S is a liquid "bio-fertilizer" of selective strain of sulphur solubilizing bacteria, *Thiobacillus thiooxidans*. This beneficial bacteria is suspended in liquid carrier at 1×10^9 bacterial cells/ml of the product.

Micro S-109

Micro S-109 is a liquid formulation of sulphur oxidizing bacteria of *thiobacillus* spp. at minimum cell concentration of 10^9 cfu/ml. Micro S-109 converts the unavailable sulphur to sulphur salts for easy uptake by plants through a process of oxidation. During this process, it helps in reclaiming the alkaline soil for normal cultivation by bringing down the high pH of the soil.

Soil with high pH acts as an impediment in making sulphur available to the plants, and also preventing proper establishment of root system. Addition of elemental sulphur varying from 200 kg to 500 kg and application of S-biofertilizer (*Thiobacillus-thiooxidans*) at the recommended rate will facilitate availability of sulphur related compound salts to the plants as nutrients. Also in the process of oxidation, it produces acid which brings down the pH to the optimal level.

In places where application of gypsum is practiced for reclamation of the soil with high pH, addition of S-biofertilizer (*Thiobacillus thiooxidans*) has added advantage in the process of reducing the pH/soil reclamation.

There are soils recorded to have high sulphur either in the form of elemental sulphur or complex compound, which are not available to the plant system. Addition of S-biofertilizer at appropriate ratios as per the dosage indicated, will be converting the elemental sulphur through the process of oxidation and produce simple sulfate or complex sulfate compound, which are water soluble and that the plant can easily absorb and utilize.

2.2.6 Silicate solubilizing biofertilizer

Silicon (Si) is an element abundantly available on earth's crust is second only to oxygen (Ehrlich 1981). It is the eighth most abundant elements in the universe. Its content in soil vary greatly and ranges from <1 to 45% by dry weight (Sommer et. al. 2006). Silica (SiO_2) content of soils varies from less than 10% to almost 100%. Silicon dioxide (SiO_2/silica) comprises of 50–70% of the soil mass. As a consequence, all plants rooting in soil contain some Si in their tissues. Soluble Si has enhanced the growth, development and yield of many plants. The content of silica in plants is equivalent to or more than the major nutrients N, P, K supplied through fertilizers. Plants may not be that much foolish to accumulate an element without any specific role in nutrition or physiology. Perhaps with the available knowledge, man is not able to pin point its exact role, requirements, etc., even though he could observe its beneficial effects on crop growth, pest and disease suppression and yield.

Silica has a surprisingly large number of functions in plants (www.somphyt.com). These are particularly as follows:

- The strengthening of epidermal cells in leaves and stems; confers non lodging.
- It is important constituent of DNA and RNA, i.e. silica deficiency decreases the synthesis of proteins and chlorophyll.
- Decreasing toxicity: Silica regulates plant's uptake of iron, manganese and aluminium. The infamous toxicity of these elements in acid soil can be counteracted with soluble silica.
- Water balance: Low silica content increases the transpiration rate (water loss through leaves) creating poor water-use efficiency.
- Improved plant growth and yield: Increased root growth in grasses, spectacular yield increases for cucumbers (1500%), 30–50% in cane and substantial increased yield in beets.
- Increased rates of photosynthesis partially due to stronger stems produce more erect leaves, which capture more sunlight
- Improved reproduction: Studies have found enhanced pollination in tomatoes and better pollen fertility in cucurbits.
- Increases resistance to pests and diseases.
- Increases the grain yield and reduces chaffiness.

Among the plants, silica concentration are found to be higher in monocotyledons than in dicotyledons; and its level increased from legumes < fruit crops < vegetables < grasses < grain crops (Thiagalingam et. al. 1977). Grasses accumulate 2–20% foliar dry weight as hydrated polymer or silica gel. Rice accumulates 4–20% silicon in straw and almost every part of rice contains this element which is not at all added exogenously as fertilizer as done with nitrogen, phosphorus and potassium – the trinity of nutrients.

Man ignored exogenous application of this element with the belief that the soil itself can sustain its supply. Unfortunately, the silica that occurs in soil is in an unavailable polymerized form and for its absorption by plants it has to be depolymerized and rendered soluble by means of biological or chemical reactions in soil. Plants absorb Si exclusively as mono silicic acid (ortho-silicic acid) by the diffusion and also by the influence of transpiration induced root absorption (mass flow). The soil silicates are the source of silica. Applications of metal silicates to rice in Japan and to sugar cane in Brazil have been extensively adapted. By looking at the amazing quantity of silicon removed from world arable soils which are estimated to the tune of 210–224 million tons annually (FAO estimate), one cannot remain silent without making any effort to render it available from *in situ* or applied through silicon fertilizer.

Silicon does not form a constituent of any cellular components, but primarily deposited on the walls of epidermis and vascular tissues conferring strength, rigidity and resistance to pests and diseases. Although no biochemical role has been positively identified in the development of plants, it has been proposed that enzyme silicon complexes formed are found to protect or regulate photosynthesis. Silicon was found to suppress the activity of certain enzymes and suppression of inverts, particularly resulted in greater supply of essential high energy precursors needed for optimum growth. It was also suggested that silica in plants filter harmful ultraviolet radiation reaching leaf surface with leaf cells, acting as 'windows', transmitting the light energy to photosynthetic mesophyll and cortical tissues beneath epidermis, than that would occur if silica were absent (Tisdale et. al. 1985).

Silicon supply increased the photoassimilation of carbon and also promoted the assimilated carbon to the panicle in rice (Takahashi and Miyake 1982). Silica also plays a role in phosphorus nutrition and there is an interrelationship with phosphorus (Silva 1971). Considering all these favorable effects, earlier workers investigated the role of exogenous application of siliceous material to rice and sugarcane and the response varied with the soil types (Padmaja and Verghese 1972; Elawad et. al. 1982).

Plants take up different quantities of silica according to their species (Russell 1973). Gallo et. al. (1974) observed that in rice, oat, rye and wheat, seed coat accumulate the most silica and grain the least. It was found that silica content of rice plant increased with the age of the crop from transplanting to harvest (Nayar et. al. 1982). The silica content of rice straw at harvest ranged from 4.8 to 13.5% in dry season and from 4.3 to 10.3% in wet season (Nayar et. al. 1982). Field trials conducted with SSB showed that this bacterium enhanced the growth, chlorophyll content, thousand grain weight, filled grains, biomass and yield in rice. Ciobanu (1961) found that 'Azotobacterin' and 'Silicabacterin' when applied simultaneously increased the yields of raw cotton by 50–34%. The beneficial effects of Si Sol B on Lucerne and maize were also shown by Vinitikova (1964).

The solubilized silicon has a larger interaction with other nutrients, particularly phosphorus. In soil system also, application of silicates released more of phosphorus (Chinnasami and Chandrasekaran 1978).

The response of crops to silicon application, particularly rice and sugarcane have been extensively investigated both in solution and soil culture by several investigations (Ota et. al. 1957; Padmaja and Verghese 1972; Dong et. al. 1981). Application of silica to rice was found to increase the grain yield under both upland and water logged conditions (Datta and shinde 1985).

Greenhouse and field experiments showed sustainable benefits of Si fertilization for rice, barley, wheat, corn, sugarcane, cucumber, tomato, citrus, and other crops (Epstein 1999).

Silicon is an element which is useful to the healthy growth of plants. It induced resistance to biotic (pest and diseases) and abiotic stress (Droght, Al, Mn, Fe toxicity alleviation), increased availability of phosphorus, increase in rigidity with reduction in drooping or lodging, improved leaf, stalk erectness, freeze resistance, and improvement in water economy.

Silicon is not considered as primary nutrient, but it is a fact that plants do accumulate silica (SIO_2) and deposit it on the walls of epidermis and vascular tissues. Soil contains 32% silica element by weight. In soil it occurs in a polymerized state as the silicate (SIO_3) minerals which are unavailable to plants. Unless depolymorized to silicic acid, it is not absorbed by the plants. Plants absorb ortho silicic acid along with water; water is lost through transpiration and silica (SIO_2) is deposited on the epidermal layers and vascular tissues in the form of silica gel.

Soil supplies some quantity of silica, but it is very low and inadequate to support fast growth of crop of plants over a localized area. Further when more of nitrogenous fertilizers are used, plants become tender and succulent; and to offset these weaknesses, more quantity of silica is necessary for crops.

There is plenty of total silicon in soil, but most of them are unavailable to plants. Silicate mineral solubilizing bacteria can dissolve silicate mineral (such as feldspars and micas) and release the elements of silicon besides potassium and phosphate.

Soil contains a variety of micro organisms, but a few are capable of solubilizing silicates. A virulent silicate solubilizing bacterium (Si SOL B) was isolated and tested on a variety of crops in different soils; it was found to release soluble silica in soils and also from silicate minerals. This bacterium is used as a biofertilizer and found to enhance the growth, suppress pests and diseases and increase the yield (Muralikannan 1996; Muralikannan and Anthoni Raj 1998).

The strain with high efficacy in silicate mineral – dissolution and production of IAA and siderophores was identified as *Rhizobium* sp. The available silicon in soil when inoculated with this strain was significantly increased by 17.3%.

The mechanism by which silicate solubilizing bacteria (SSB) suppresses various fungal diseases in plants is contributed to the resistance provided by accumulation of silica on cell wall which unable the pathogens in penetrating host tissue as demonstrated in case of pyricularia aryzoe. A SSB strain Q-12 isolated from the surface of feldspar and identified as Bacillus globisporus Q-12 has potassium solubilizing activity also. The efficiency of the SSB on plant growth stimulation and biocontrol benefits to pyricularia oryzae in rice due to Bacillus mucilaginosus is well-documented.

2.2.7 Silica solubilizing bacillus sp.

Silica unlike other essential element in plants has a function that is mechanical rather than physiological. This characteristic of Silica function explains why Si effects are easily observed in plants that accumulate Silica to a certain extent and why Si effect are more obvious under biotic or abiotic stress.

Silicates make plant cell walls thicker and stronger while increasing the size of the vascular system of the plant. Thicker cell wall makes plant stronger in all aspects. Enlarged vascular system aids in the plant, taking up more water and nutrients resulting in bigger, healthier, and higher yielding plant.

Si Sol B™ is a biological fertilizer based on a selected strain of naturally occurring beneficial bacteria of the Bacillus genus isolated from a granite quarry. Si Sol B™ is used as effective soil inoculants. It solubilizes silica and provides the plant with strength to tolerate biotic and abiotic stresses and improves its resistance to pest and disease attack. Si Sio B^R contains spores of the Bacillus spp. It is formulated as Wettable Powder with CFU count of 1×10^8/g.

Si SOL B™ containing the microbe produces organic acids as part of its metabolism that has a dual role in silicate weathering. They supply H+ ions to the medium and promote hydrolysis and the organic acids like citric acid, oxalic acid, keto acids and hydroxyl carbolic acids which form complexes with cations and make the silica available to the plant in an assumable form.

The solubilization of silicates is demonstrated using kaolin and quartz sand as model substances. The process is more effective in the alkine than in the acid pH range. In the acid medium, oxalic acid showed maximum acidity and a tendency to form complex structures, especially with aluminum and was most effective in leaching. The highest leaching activity was observed in the case of Thiobacillus thiooxidans, whereas the heterotrophic microorganisms (among them Bacillus mucilaginosus) did not exercise a solubilizing effect on the silicates.

Fortis-Silicate solubilizing Bacterium (SSB), a Bacillus sp., is a unique bacterium. Application of Fortis-SSB to soil not only increases Si status of plants, but also improves status of P and K of plants and there by rendering plants more resistant to pests and disease attack.

Silica solubilizing bacteria has been found to solubilize soil silicates releasing silicic acid which is absorbed by plants. Incoulation of Fortis – SSB to rice was found to increase silica levels in soil and plants, augmented growth and yield.

2.2.8 Decomposting cultures

Composting as a recognized practice dates to at least the early Roman Empire since Pliny the Elder (AD 23–79). Traditionally, composting involved piling organic materials until the next planting season at which time the materials would have decayed enough to be ready for use in the soil. The advantage of this method is that little working time or effort is required from the composter and it fits in naturally with agricultural practices in temperate climates.

Composting was somewhat modernized in the beginning of the 1920s in Europe as a tool for organic farming (Heckman 2006). The first industrial station for the transformation of urban organic materials into compost was set up in Wels/Austria in the year 1921 (Welser Anzeiger 1921). Early frequent citations for propounding composting within farming came from the Germans namely, Rudolf Steiner, founder of a farming method called biodynamics, and Annie France-Harrar, who was appointed on behalf of the government in Mexico and supported the country (1950–1958) to set up a large humus organization in the fight against erosion and soil degradation. In the English-speaking world it was Sir Albert Howard who worked extensively in India on sustainable practices and Lady Eve Balfour who was a huge proponent of composting. Composting was imported to America by various followers of these early European movements by the likes of J. I. Rodale (founder of Rodale Organic Gardening), E. E. Pfeiffer (who developed scientific practices in biodynamic farming), Paul Keene (founder of Walnut Acres in Pennsylvania), and Scott and Helen Nearing (who inspired the back-to-the-land movement of the 1960s).

There are many modern proponents of rapid composting who attempt to correct few of the perceived problems associated with traditional slow composting. Many advocate that compost can be made in 2–3 weeks (http://vric.ucdavis.edu). Many such short processes involve a few changes to traditional methods, including smaller, more homogenized pieces in the compost, controlling carbon-to-nitrogen ration (C:N) at 30 to 1 or less, and monitoring the moisture level more carefully.

As concern about landfill space increases, worldwide interest in recycling by means of composting is growing, since composting is a process for converting decomposable organic materials into useful stable products (www.stormcon. com). Composting is one of the only ways to revitalize soil vitality due to phosphorus depletion in soil (www.netwrx.com). Industrial-scale composting in the form of in-vessel composting, aerated static pile composting and anaerobic digestion takes place in most of the Western countries now; and in many areas, it is mandated by law. There are process and product guidelines in Europe that date to the early 1980s (Germany, The Netherlands, Switzerland) and only more recently in the United Kingdom and the United States. Compost is regulated in Canada and Australia as well.

Industrial composting systems are increasingly being installed as a waste management alternative to landfills, along with other advanced waste processing systems. Mechanical sorting of mixed waste streams combined with anaerobic digestion or in-vessel composting is called mechanical biological treatment and are increasingly being used in developed countries due to regulations controlling the amount of organic matter allowed in landfills. Treating biodegradable waste before it enters a landfill reduces global warming from fugitive methane; untreated waste breaks down anaerobically in a landfill, producing landfill gas that contains methane, a potent greenhouse gas.

Large-scale composting systems are used by many urban areas around the world. The World's largest MSW co-composter is the Edmonton Composting Facility in Edmonton, Alberta and Canada, which turns 220,000 tonnes of residential solid waste and 22,500 dry tonnes of biosolids per year into 80,000 tonnes of compost. The facility is 38,690 m² (416,500 ft²) and the operating structure is the largest stainless steel building in North America, the size of 14 NHL rinks (www.edmonton.ca).

Another large MSW composter is the Lahore Composting Facility in Lahore, Pakistan, which has a capacity to convert 1000 tonnes of municipal solid waste per day into compost. It also has a capacity to convert substantial portion of the intake into Refuse-derived Fuel (RDF) materials for further combustion use in several energy consuming industries across Pakistan; for e.g., in cement manufacturing companies, where it is used to heat up the cement kiln systems. This project has also been approved by the Executive Board of the United Nations Framework Convention on Climate Change (UNFCCC) for reduction of emission of methane gas into the climate and has been registered with a capacity of reducing 108,686 metric tonnes CO_2 equivalent per annum (http://cdm.unfccc.int).

Urban solid waste management is considered to be one of the most serious environmental problems confronting urban areas in developing countries.

Major portions of the Municipal Solid Waste (MSW) are highly degradable organic matter. If the organic matter in MSW is not handled properly, it will be wastage of resource and it will contribute in forming heaps of waste and environmental problems at landfill site.

Composting is the process in which organic substances are reduced from large volumes of rapidly decomposable materials to small volumes of materials which continue to decompose slowly. Composting is an ancient technology, practiced today at every scale from the backyard compost pile to large commercial operations. Composting reduces the burden on government authorities for transportation and disposal of MSW at landfill sites. Compost is easier to handle than manure and other raw organic materials; it stores well and is odor free, added to farms or gardens to improve soil structure, texture, aeration and water retention and to increase soil productivity. Mixing compost with soil also contributes to erosion control, soil fertility, proper pH balance, and healthy root development in plants.

Composting is a self-regulating, thermophilie, aerobic process, involving the microbial conversion of biodegradable organic wastes to stable humus by indigenous micro-flora including bacteria, fungi, actinomycetes, and macro-fauna like earthworms. By biocomposting the organics, saving occur in waste transportation costs. Lesser contamination of recyclables lead to better reuse and health hazard due to putrefying biomass is mitigated.

Effective microorganism (EM) solution, leachate from landfill site, biosanitizer, earthworms, decomposting culture, and its different combinations are used for composting. Study reveals that leachate addition to composting process gives high degradation rate (75.78%) in 55 days, but nutrient content of this compost as C/N ration was less than the other methods. Degradation of organic matter in combination of vermin-composting and EM solution are faster within 40 days as compared to only single addition of decomposting culture, and also nutrient content is more compare to other methods.

Both bacteria and fungi are present and active in a typical composting process. The major bacterial groups in the beginning of the composting process are mesophilic organic acid producing bacteria such as *Lactobacillus* spp. and *Acetobacter* spp. Later at a thermophilic stage, gram-positive bacteria such as Bacillus spp. and Actinobacteria become dominant. However, the most efficient composting process is achieved by mixed communities of bacteria and fungi.

Among bacteria, bacillus spp. *Thiobacillus, Azotobacter, Caryophanon, Cellulomonas* and *Sporosarcina*.; while among *actinomycetes, streptomyces* spp., *Thermomonospora* spp., *saccharomonospora* spp., *Actinomadura* spp., *Rhodococcus* spp. *and faenia* spp.; and among fungi, *Aspergillus* spp., *Humicola* spp., *Torula* spp., *Penicillium* spp., *Thermomyces, Chaetomium, and Mucor* are used as composting microbial cultures.

Consortiums of composting cultures are also available. RSCLBIO-D is a consortium of live aerobic microorganisms, specially developed for composting or degrading organic wastes like sugar factory press mud, distillery spent wash, cow dung, poultry manure, coir pith, sugarcane trash, bagasse, city garbage and other agricultural wastes. This eco-friendly microbial-compost starter culture helps to accelerate the organic matter decomposition with increased humidification and shortened composting time. One to two kg of RSCL BIO-D/MT of organic wastes is mixed with water at 1 kg/10 l of water, aerates for 24 h and applied on the compost heaps or compost pits. RSCL BIO-D can be stored for 1 year in a cool and dry place with temperature less than 30°C and should be kept away from direct sunlight. While traditional composting procedures take as long as 4–8 months to produce finished compost, rapid composting methods offer possibilities for reducing the processing period up to three weeks.

The world wide need to restore the productivity and humus forming ability of infertile or overburdened soils requires the application of special soil inoculants, organic composts and/or fertilizers. The large amounts of unused or underused biomass resources such as agricultural and forestry residues, industrial and municipal organic waste, may serve as raw material for the preparation of compost, specially designed for the need of a particular crop. For composting, special microbial starter cultures are prepared by anaerobic or aerobic submerged fermentation of 6–8 different microorganisms/bacteria including Actinomycetes. The microbes are grown to 10^9–10^{10} CFU/ml and mixed in the appropriate proportion to suit the composting process, directed to a particular plant. About 5–10% v/v of the mixed microbial culture is added to the selected organic carrier and composted anaerobically or aerobically at a determined optimal temperature, pH, for 25–60 days. In such directed compost, the desired added microbes are the predominant species. Directed composts are prepared for field crops, vegetables, ornamentals and lawns, using fruit and winery processing residues, municipal sewage sludge, agricultural and forestry residues, etc. Such composts are now marketed as commercial products. Trichoderma harzianum, a biological plant protection agent, is now successfully incorporated into such composts to have additional benefits of soil-borne disease management.

3.1   Introduction

Most of the cultivated tropical soils have good population of rhizobia, capable of nodulating legumes. Presence of nodules should not be taken as an index of establishment of an efficient nitrogen fixing system. Competitive and efficient strains of *Rhizobium* used for inoculation will ensure maximum nitrogen fixation. Absence of effective *Rhizobium* strains is responsible for poor nodulation encountered in many areas. Absence of suitable rhizobia, deficiency or toxicity of a nutrient, unfavorable soil conditions, excess water logging, unsuitable pH, predators and pests are the other factors which indirectly influence the potentiality of *Rhizobium* strains.

Drought and temperature stress is known to affect the growth of legume host and nitrogen fixation. Nodule formation and nitrogen fixation are influenced by the rhizosphere temperature. The capacity to tolerate elevated temperature varies considerably among different strains of rhizobia and host species. Therefore, the ability to tolerate higher temperature will be a very desirable trait for the selection of *Rhizobium* strains in a tropical country like India as this would help in build up and carryover of strains introduced into soil from one season to another. The *Rhizobium* strains are known to have significant potential of nitrogen fixation ranging from 50 to 300 kg/ha per season, depending on the crops. Further, legumes often require specific strains of *Rhizobium* for maximum nitrogen fixation and nodule formation.

Singleton et. al. (1992) noted that rhizobial inoculation is underutilized in tropical agriculture and the size of populations of indigenous rhizobia has a major impact on the establishment and symbiotic performance of inoculant rhizobia.

3.2   Isolation of *Rhizobium*

It is not always necessary to excavate the whole plant to retrieve nodules. Careful removal of soil around the root crown to locate young lateral roots having nodules, followed by removal of the roots with a knife often yields best results. If this method proves unsuccessful (it depends very much on the plant species involved), then deeper excavation of the root system may be necessary. Use a digging implement to remove soil which is approximately 25 cm in diameter containing roots to a depth of 10–15 cm. Exposure of the root system is greatly facilitated by immersion of the root in water and allowing the soil to fall away. The soil and roots can be washed in a sieve (1–2 mm openings) either by "dunking" or with running water to expose the root nodules. Exposed nodules can then be collected with forceps.

The location of nodules on a root system is dependent on plant species. For tap-rooting plants such as *Trifolium subterraneum*, nodules can be found within 2–4 cm of the crown on tap and lateral roots; whereas nodules on stoloniferous legumes are usually concentrated superficially within 1–2 cm of the surface, attached to adventitious roots. In most stylosanthes species, nodules are more or less equally distributed along the length of the primary and lateral roots in the top 10–15 cm soil strata.

Do not expect to locate nodules on the main rootstock of perennial species. Nodules of most legumes have a definite life span; and a maturing root that may have borne nodules when younger, no longer has the anatomy that would permit re-infection and nodulation. Similarly, do not expect to find nodules on a plant that has been pulled out of the ground by force. In most of the species, the connection between the nodule and root is extremely tenuous.

Sample only fresh and firm nodules, and avoid taking samples of those that are damaged or decaying. A sample of at least 20–25 nodules per plant is recommended. The reason for the large sample is that not all strains of rhizobia forming nodules on the host are effective in nitrogen fixation; thus, it may be necessary to make isolations from a number of nodules to ensure success in obtaining effective isolates. It is necessary to select 5–10 nodules from each sample to be processed, leaving the remainder as backup for further attempts if the initial isolations are unsuccessful. This is especially important where nodules have been obtained during long-term trips and revisiting the site for further collection is not possible. All nodules from a single plant represent one unit of collected material and may be stored in the same container. Nodules from different plants should not be combined because they may represent different soil environments even if only several meters apart and in extreme cases (e.g., misidentification) could lead to a mixture of nodules from different species.

3.2.1 Materials

For isolation of *Rhizobium* from leguminous plant, the requirements are: nodule of leguminous plant (sunhemp/soybean/French bean/gram, etc.), mercuric chloride solution (1:1000), alcohol (75%), sterile petridishes, sterile water and yeast extract mannitol agar (Waksman's medium No. 79) medium with Congo red dye.

Yeast Extract Mannitol Agar Medium

Mannitol	10 g	K_2HPO_4	0.5 g
$MgSO_4H_2O$	0.2 g	NaCI	0.1 g
Yeast Extract	1.0 g	$CaCo_3$	3.0 g
Agar	20.0 g	Congo red (1%)	2.5 ml
Distilled water	1000 ml	pH	7.0

Agar and Congo red used for solid medium

10 ml of sterile 1:400 aqueous solution of Congo red per liter of YEMA medium is added to the medium for preparing Congo red yeast extract mannitol agar medium.

3.2.2 Procedure

(1) Uproot the plant carefully that has attained the flowering stage and observe for the presence of root nodules.

(2) Wash the root system thoroughly with tap water to remove adhering soil.

(3) Record the observations regarding the nodulation pattern particularly, number of nodules on tap root system, on lateral root system, total number of nodules, biomass of the nodules (which can be done either by fresh weight, or by immersing the nodules in measuring cylinder containing water and finding out how much quantity of water is raised), color of nodules, i.e. number of red, pink, white, or brown nodules and size of nodules.

(4) For isolation, select only those nodules which are bigger in size, located on the tap root system, are pink in color and succulent.

(5) Detach such nodules with the help of a blade/forceps in such a way that they should not be injured. However, it should be cut along with the small rootlets.

(6) Disinfect these nodules for 3–5 minutes in $HgCl_2$ (1:1000).

(7) Wash these nodules in three changes of sterile water.

(8) Transfer one or two nodules in an empty sterile petridish. Add 1 ml sterile water and crush the nodules with sterile scalpel. After crushing it, one can find that a big bacterial mass oozes out.

(9) In few other sterile petriplates add 1 ml sterile water and transfer loopful culture from crushed biomass to each plate. Rotate the plates in order to adhere the bacteria to the internal bottom of plate. Pour all the plates with Congo red yeast extract mannitol agar medium and make the homogenous mixing of bacterial suspension with the medium.

(10) Incubate the plates at 28 ± 1°C.

(11) After incubation of these plates for a week at 28 ± 1°C temperature, pick up the colonies which are white raised, shinning, translucent with the smooth edge, on the slants of YEMA medium. The colonies that are deep red in color can be of common contaminant like *Agrobacterium tumefaciens* which are normally associated with legume root nodules.

(12) Transfer the representative colonies on yeast extract mannitol agar slants for further studies of nodulation.

3.2.3 Observations

Examine the plates for 4–5 days after incubation for the development of Rhizobial colonies. The colonies are raised, wet, shining and somewhat translucent with smooth edges on the surface of yeast extract mannitol agar. On Congo red agar, rhizobia will produce whitish colonies; whereas the common contaminant like *Agrobacterium tumefaciens* will produce deep red color (Figures 1 and 2).

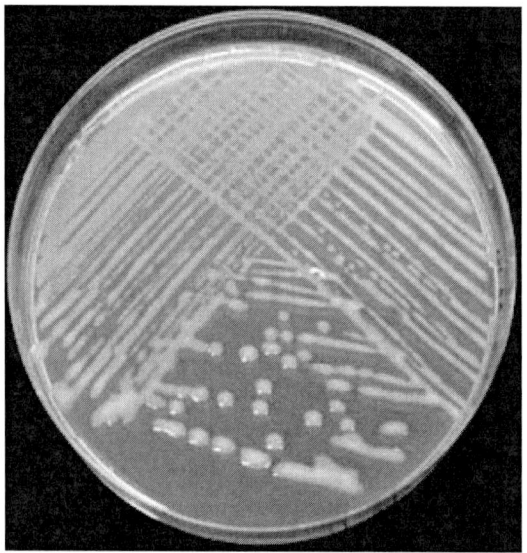

Figure 3.1 Colonies of *Rhizobium.*

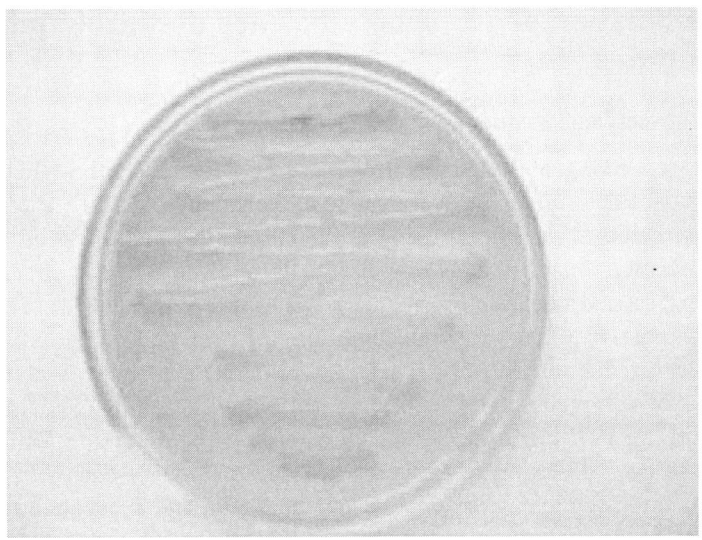

Figure 3.2 Colonies of contaminant.

3.3   Cultural tests to distinguish Rhizobia from contaminants

Rhizobium colonies formed on Congo red YEMA medium are white, translucent, glistering, elevated and comparatively small. Common contaminants are Agrobacterium colonies on YEMA media.

(1) Congo red test: Most of the Rhizobia including *Bradyrhizobium* lack the ability to absorb Congo red from YEMA medium (containing 0.0025% final concentration of this dye). *Agrobacterium* absorbs Congo red readily and becomes pink initially and then turn deep black.

(2) Growth in alkaline medium: *Agrobacterium radiobacter* could grow even at pH 11 whereas, *Rhizobium* does not grow. Add 1.6% BTB indicator (1 ml/lit) in YEMA. Adjust the pH to 11. Observe the growth up to 15 days. The change in color of BTB indicator which turns from blue to green indicates *Agrobacterium*.

(3) Growth in glucose-peptone agar: All rhizobia, except a few strains of rhizobia, show no or little growth on glucose-peptone agar, whereas *Agrobacterium* grows well.

3.4 Estimation of 'N' fixing efficiency of Rhizobium cultures

(1)   By Acetylene Reduction Assay (ARA)

The nitrogen fixing ability, i.e. nitrogenase activity of all the Rhizobium isolates can be determined by acetylene reduction assay (ARA). The ethylene (C_2H_4) formation is measured using a gas chromatograph equipped with a flame ionization detector and Porapak T column. The ARA for free-living rhizobia is carried out as prescribed by Kaneshiro et. al. (1978). The free living rhizobia requires stringent conditions of growth for the expression of acetylene reduction activity. The specific defined medium used for the induction of nitrogenase in free-living *Rhizobium* is solid CS 7 medium (Gibson et. al. 1976). In brief, the slants are prepared by taking 5 ml CS 7 medium in each of 15 ml test tubes. A loopful culture of each *Rhizobium* isolate in three replicates is streaked on the slants and incubated at $28 \pm 2°C$ for 8 days. The cotton plugs are replaced with air-tight rubber serum stoppers, 10% of air in test tube is removed and replaced with equal volume of acetylene gas using gas-tight syringe. The tubes are incubated at $28 \pm 2°C$ for 24 hours. After completion of incubation period, 1 ml of gas sample from test tube is injected into the pre-conditioned gas chromatograph for ethylene estimation. The protein concentration is determined by a modified Lowry's method (Lowry et. al. 1951) with bovine serum albumin (BSA) as standards. The nitrogenase activity is calculated by using following formula:

$$\text{nmole } C_2H_4 \text{ produced, Mg protein}^{-1}, Hr^{-1} = \frac{Ce \times Ps \times Va \times As \times 60}{Pstd \times Vs \times Astd \times T \times P}$$

Where:

Ce	=	Concentration of ethylene in standard (in nmoles)
Ps	=	Peak height for sample (in cm)
Va	=	Volume of assay tube gas phage (in ml)
As	=	Attenuation used for sample
Pstd	=	Peak height for standard ethylene (in cm)
Vs	=	Volume of gas sample injected for analysis (in ml)
Astd	=	Attenuation used for standard ethylene
T	=	Time for incubation (in min)
P	=	Protein content of bacterial growth on slant (in mg)

(2)   By Kjeldahl method

Prepare yeast extract mannitol broth to obtain growth of Rhizobium. After 12 days growth at 28°C, test the contents of the flask for purity by streaking on fresh medium and concentrate the bacterial broth over a water bath (50–60°C) to dryness. Wash the dried culture and take it as a sample. The content of the flask in inoculated control series should be processed in the similar manner.

(1) *Reagents*: mercuric oxide

(2) *Sulphuric acid*: 93–93 percent, N-free

(3) *Digestion mixture*: Mix copper sulphate and potassium sulphate in the ratio 1:10 and grind them into fine powder.

(4) *Sodium hydroxide pellets or solution, N-free*: For solution , dissolve about 450 g of Sodium hydroxide in water, cool and dilute to 1 liter (sp. gr of the solution should be at least 1.36).

(5) *Zink granules*: reagent grade.

(6) *Indicators*:

    a.   Methyl red indicator: Dissolve 1 g of methyl red in 200 ml ethanol

       i. Mixed indicator: Prepare mixed indicator by dissolving 0.5 g of methyl red and 0.2 g methyl blue in 500 ml of ethanol.

(7) Hydrochloric/sulphuric acid: Standard solution 0.5 or 0.1 N when amount of nitrogen is small.

(8) Sodium hydroxide standard solution: 0.1 N (or other specified concentration).

Note: Ratio of salt to acid (w/v) should be about 1:1 at the end of digestion for proper temperature control. Digestion may be incomplete at a lower ratio and nitrogen may be lost at higher ration. Each gram of fat consumes 10 ml of sulphuric acid and each gram of carbohydrate 4.0ml of sulphuric acid during digestion.

3.4.1 Apparatus

(1) For digestion: Use Kjeldahl's flask of hard moderately thick well-annealed glass with total capacity of approximately 500–800 ml. Conduct digestion over heating device adjusted to bring 250 ml of water at 25°C to rolling boil in about 5 minutes. To test the heaters, preheat for 10 minutes in case of gas burners and for 30 minutes in case of electric heaters. Add 3–4 boiling chips to prevent super heating.

(2) For distillation: Use 500–800 ml Kjeldahl's flask fitted with rubber stopper through which passes the lower end of an efficient scrubber bulb or trap to prevent mechanical carryover of sodium hydroxide during distillation. Connect the upper end of the bulb tube to a condenser by rubber tubing. Trap the outlet of the condenser in such a way as to ensure absorption of ammonia distilled over with the receiver.

3.4.2 Procedure

Place 0.25 g of the sample in the digestion flask. Add 0.7 gm mercuric oxide, 15 gm potassium sulphate digestion mixture, followed by 25ml sulphuric acid. Shake, let stand for about 30 minutes and then heat carefully until frothing ceases. Boil briskly until the solution clears and continue boiling further for 90 minutes. Cool, add about 200 ml of water, cool to room temperature and add a few Zink granules.

Tilt the flask and carefully add 50 ml of sodium hydroxide solution without agitation. Immediately connect flask to the distillation bulb on the condenser of which the tips are immersed in 50 ml of standard 0.1N acid in the receiving flask. Rotate the digestion flask carefully to mix the contents. Heat until 150 ml of the distillate collects and titrates excess acid with 0.1 N base using methyl red or mixed indicator. Carry out blank determination on reagents.

Note: Check the ammonia recording periodically by using inorganic nitrogen control; for example, ammonium sulphate.

3.4.3 Calculation

(1)   Nitrogen content, percent by mass

$$= \frac{\text{Milliliters of 0.1 N acid for sample} - \text{milliliters of 0.1 N acid for blank} \times 0.14}{\text{Mass of sample taken}}$$

(2)   Total nitrogen in culture = Total dry mass of sample × percent nitrogen.

Take 1.0 g of accurately weighted sample each from the inoculated series and from the controls. Put them separately in 250 ml volumetric flask, add 150 ml water, mix the content and increase the volume to 250 ml by adding water in it. Shake well for 5 minutes and centrifuge for 15 minutes at 10000rev/min. Estimate glucose in the supernatant in triplicate. The difference between two provides the data of actual amount of glucose consumed; calculate the amount of nitrogen fixed per gram of sucrose consumed.

1.   **Determination of Glucose**: From the supernatant, draw the suitable aliquots and estimate reducing sugars (glucose) as follows:

Reagents

Soxhlet modification of fehling solution: Prepare it by mixing equal volumes of *solution A* and *solution B* immediately before using.

Solution A:

Dissolve 34.639 g of copper sulphate crystals (Cu 4. 5H20) in water, dilute to 500 ml and filter through glass cool or filter paper.

Solution B:

Dissolve 173 g of potassium sodium tatrate and 50 g of sodium hydroxide in water and dilute to 500 ml. Let the solution stand for a day and filter.

Hydrochloric acid — sp gr 1:18 at 20°C (approximately 12 N)

2. Standardization of copper sulphate solution:

Using separate pipettes transport accurately 5 ml of solution A and 5 ml of solution B into a conical flask of 250 ml capacity. Heat this mixture to boiling point on an asbestos gauze and add standard invert sugar solution from a burette, about 1ml less than the expected volume which will reduce the Fehling solution completely (about 48 ml). Add 1ml of methylene-blue indicator while keeping the solution boiling. Complete the titration within 3 minutes; the end point being indicated by the change of color from blue to red. From the volume of invert sugar solution used, calculate the strength(s) of the copper sulphate solution by multiplying titre value by 0.001 (mg/ml of the standard invert sugar solution). This would give the quantity of the invert sugar required to reduce the copper in 5 ml of copper sulphate solution.

3. Standard invert sugar solution:

Weight accurately 0.95 g of sucrose and dissolve it in 500 ml of water. Add 32 ml concentrated hydrochloric acid, boil gently for 30 minutes and keep aside for 24 hours. Neutralize with sodium carbonate and the make the final volume to 1000 ml; 50 ml of this solution contains 0.5 g of invert sugar.

Methylene blue indicator–0.2% in water

(i) Procedure

Place about 1 g (M), accurately weighed, of the prepared sample of Al in volumetric flask and dilute it with about 150 ml of water. Mix the contents of the flask thoroughly and increase the volume up to 250 ml with water. Using separate pipettes, take accurately 5ml each of solution A and B in porcelain dish. Add about 12 ml of Al solution from a burette and heat to boiling point over asbestos gauze. Add 1 ml of methylene blue indicator while keeping the solution boiling. Complete the titration within 3 minutes; the end point being indicated by change of color from blue to red. Note the volume (H) in ml of AI solution required for the titration.

(ii) Calculation

$$\text{Total reducing sugar, percent by mass} = \frac{250 \times 100 \times S}{H \times M}$$

Where:

S = Strength of copper solution

H = Volume in ml of Al solution required for titration

M = Mass in g of Al taken for the test

4. **Determination of sucrose**

   (i) **Procedure:** To 100 ml of the stock Al solution, add 1 ml of concentrated hydrochloric acid and heat the solution to near boiling point. Keep aside for overnight. Neutralize this solution with sodium carbonate and determine the total reducing sugar as described.

   (ii) **Calculation:**

   (a) Sucrose, percent by mass = (reducing sugar after inversion, percent by mass) –(reducing sugar before inversion, percent by mass) × 0.95

   (b) Nitrogen, mg per gram of sucrose consumed = $2 (a - b) - c$

Where:

a = Initial quantity of sucrose taken for the test

b = Mass of sucrose as calculated in (a)

c = Amount of nitrogen fixed per gram of glucose

4.1 Introduction

Among various nitrogen-fixing bacteria, *Azotobacter* is one of the most important non-symbiotic nitrogen-fixing bacterium. It is a free-living bacterium and is found in the rhizosphere of cereal crops. The soils rich in organic matter harbor more population of *Azotobacter*. It is an aerobic, gram-ve, short rod and heterotrophic bacterium. *Azotobactor* not only provides the nitrogen, but also produces a variety of growth promoting substances. Some of these are indole acetic acid (IAA), gibberallins, vitamin-B and antifungal substances. Another important characteristic of *Azotobactor* associated with crop is the excretion of ammonia in the rhizosphere in presence of root exudates. These are better competitors for higher survival rate in soil as compared to the non-excreting strains. *Azotobacter* serves as a broad spectrum inoculant used for various crops like wheat, barley, maize, paddy, jawar, oat, sugarcane, sunflower, mustard, sesamum, linseed, tea, coffee and all types of forest fruit and flower plants. *Azotobacter* also causes mineralization of fixed phosphate in soil and thus increases the uptake of P in plants. Crops receiving *Azotobacter* inoculation along with moderate levels of fertilizer nitrogen gives similar grain yields as the crops receiving higher doses of fertilizer.

Its isolation from the rhizosphere soil of any of the cereal crops can be made by two methods, viz. (1) Soil dilution and pour plate technique and (2) Enrichment culture technique.

For isolation of *Azotobacter*, either Jensen's or Ashby's medium can be used. These media are devoid of nitrogen source.

4.2 Isolation of *Azotobacter* from soil on selective Jonsen's agar medium

4.2.1 By soil dilution and pour plate technique

Azotobacter population is abundant in the rhizosphere of cereal crops. On dilution of rhizosphere soil, *Azotobacter* cells get separated and subsequently on plating on nitrogen free Jensen's agar or Ashby's medium, individual cells develop into colony. Such colonies developed from individual cell yield pure culture of *Azotobacter* which can be maintained on Jensen's agar slants.

Materials required

Rhizosphere soil of jowar/bajra/wheat/maize, petri plates, water blanks, pipettes, Jensen's agar or Ashby's agar, slants of these media, incubator, etc.

Ashby's Medium

Mannitol	10.0 g	K_2HPO_4	0.2 g
$MgSO_47H_20$	0.2 g	NaCI	0.2 g
K_2SO_4	0.29 g	K_2SO_4	0.1 g
Agar	15.0 g	pH	7.0
Distilled Water	1000 ml		

Jensen's Medium

Sucrose	20.0 g	K_2HPO_4	1.0 g
$MgSO_47H_2O$	0.5 g	NaCI	0.5 g
$FeSO_4$	0.1 g	$CaCO_3$	2.0 g
Agar	15.0 g	Distilled Water	1000 ml
pH	7.0		

Procedure

(1) Suspend 10 g of rhizosphere soil sample in 90 ml sterile water and mix thoroughly.

(2) Prepare 10 fold dilutions of the above suspension from 10^{-1} to 10^{-6}, using separate water blanks.

(3) Transfer 1 ml aliquot of the appropriate dilution to sterile petri dishes, separately.

(4) Pour in petri dishes the solidifiable Jensen's agar or Ashby's agar having temperature of 45°C.

(5) Mix the contents of the plates carefully by rotating the plates gently on working platform of Laminar-Air Flow cabinet.

(6) Allow the medium to solidify and incubate the plates at 28°C (± 2°C) in an incubator.

(7) Observe for the development of soft, flat, milky and mucoid colonies of *Azotobacter* after 3 days of incubation.

(8) Transfer typical *Azotobacter* colonies on the slants of Ashby's or Jensen's agar.

Observations:

(1) Observe the growth of *Azotobacter* by gram staining under microscope.

(2) Record dilution wise colony count and conclude the abundance of *Azotobacter* population in a given soil sample.

(3) Note down whether there is pigmentation produced by older *Azotobacter* culture in a slant or plate.

4.2.2 By enrichment culture technique

In enrichment culture technique, the nutrient medium is enriched with defined chemical components which are necessary for a particular organism to grow and show its predominance by its ability to grow more rapidly on this medium than others. *Azotobacter* being N_2 fixer, can use atmospheric N_2 and thus for its isolation the nutrient medium that is free of combined nitrogen (but contains other minerals and carbon source) is used. Jensen's or Ashby's medium is most suitable for the isolation of *Azotobacter* by enrichment culture technique.

Material required

Rhizosphere soil of jowar/bajra/wheat/maize, 250 ml conical flasks, pipettes, Jensen's agar or Ashby's agar, slants of these media, incubator, etc.

Procedure

(1) Collect rhizosphere soil of any cereal crop growing profusely.

(2) Add 0.5 g of rhizosphere soil separately to 3 to 4 conical flasks containing 100 ml of sterile Jensen's broth and shake well to mix the soil with broth.

(3) Incubate the flasks at 28°C (± 2°C) for a week.

(4) After incubation, transfer 1–2 ml of growth suspension from each of these flasks separately to another flasks containing Jensen's liquid medium.

(5) Incubate newly transferred flasks at 28°C (± 2°C) for a week.

(6) Similarly, make 2–3 more transfers and observe for the growth of *Azotobacter* obtained from the finally transferred flasks.

(7) Transfer loopful of growth on the slants of Jensen's agar.

(8) Examine the growth of *Azotobacter* under microscope for cell-shape, motility, Gram reaction, etc.

Observations

(1) Record the observations of common contaminants coming up along with *Azotobacter* in first few transfers.

(2) Note down whether more dense and uniform layer/pellicle of *Azotobacter* is formed or otherwise in the finally transferred flasks.

(3) Examine under microscope and record the observations on Gram reaction, size, shape, and motility of *Azotobacter* isolate.

4.3   Cultural test to distinguish *Azotobacter* from contaminant

On Jensen's or Ashby's agar medium, *Azotobacter* colonies are raised; slimy (Fig. 3), aged cultures show yellowish-brown/black coloration due to pigment production (Fig. 4).

On Jensen's broth, a thin layer of Pellicle is formed by *Azotobacter* (Fig. 5).

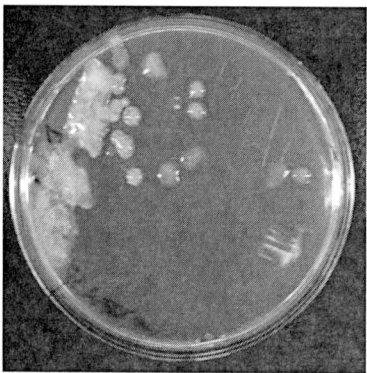

Figure 4.1 Azotobacter colonies on Asbby's.

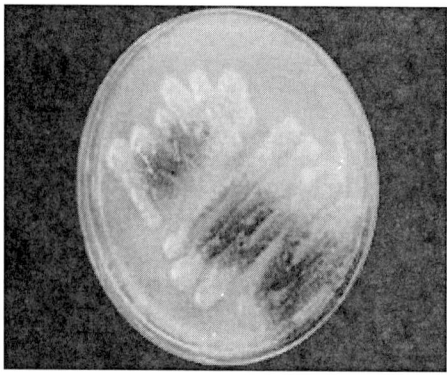

Figure 4.2 Age cultured of *Azotobacter* medium showing discoloration.

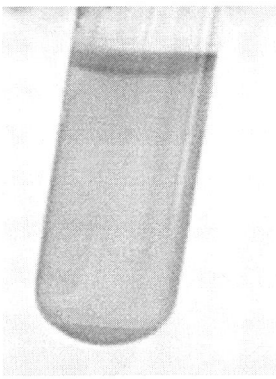

Figure 4.3 Thin layer Pellicle of *Azotobacter* on Jensen's broth.

Azotobacter produces small (0.5–1.0 mm diameter) cream opaque colonies on milk agar medium (Grey 1953). Three to four days old *Azotobacter* colonies appear as flat, soft, milky and mucoid on agar plate (Mishustin and Shilinikora 1972).

Colonies of *Azotobacter* chroococcum are generally 2–6 mm in diameter, opaque, entire, low convex, viscid, glistening and smooth. Variant colony forms may arise through desiccation in the quantity of extra cellular polysaccharides produced (Tchan and New 1986).

4.4 Test for Nitrogen fixation in pure culture of *Azotobacter*

The *Azotobacter* strain should be capable of fixing at least 10 mg of nitrogen per g of sucrose consumption. Efficiency of Nitrogen fixation of *Azotobacter* is determined by Kjeldahl Method.

4.4.1 Pure culture medium

Prepare Jensen's medium for growth of pure culture of *Azotobacter chroococcum*.

Procedure

Select the *Azotobacter* colony from pure culture. Use this pure culture for inoculating the broth for the nitrogen fixation. For this purpose, take 50 ml aliquots of broth in 250 ml conical flask for inoculation. After 12 days growth at 28°C, test the contents of the flask for purity by streaking in fresh medium and concentrating over a water bath (50–60°C) to dryness. Wash the dried culture and take it as a sample, the contents of the flask in inoculated control series should be processed in a similar manner.

The nitrogen can be estimated as per procedures described in earlier chapter for Nitrogen fixation.

5.1 Introduction

Azospirillum, being micro-aerophilic in nature and inhabiting in the upper cortex region of roots, can be easily isolated by inoculating surface sterilized root-bits into semi-solid NFb-medium. This medium supplies malate as a carbon source which is preferred by *Azospirillum*. Thin and white pellicle of *Azospirillum* develops below the surface of this medium, which can be used for transferring the growth of the bacterium to the slants of the same medium and pure culture of *Azospirillum* can be obtained.

Sodium malate medium/NFB. Medium for *Azospirillum*:

K_2HPO_4	0.5 g	Malic acid	5.0 g
$MgSO_4 \ 7H_2O$	0.1 g	Sodium chloride	0.02 g
Sodium molybdate	0.002 g	$MnSO_4$	0.01 g
KH_2PO_4	0.4 g	$FeSO_4 \ 7H_2O$	0.05 g
KOH	4.0 g	Bromoethymol blue	2.0 ml
	(5% alcoholic solution)		
$CaCl_2$	0.01 g	Distilled water	−1000 ml
Agar	1.75 g	pH	−6.8

5.2 Isolation of *Azospirillum* from soil

5.2.1 Procedure

(1) Take 1 g rhizosphere soil sample of C_4 grasses (eg., Haryali) and add in 10 ml sterile water blank, mix and shake thoroughly. This will give 10^{-1} dilution of original soil sample. Subsequently, prepare 10-fold dilution up to 10^{-7}.

(2) Take one ml aliquot for dilution (10^{-2}, 10^{-3}, 10^{-4}, 10^{-5} and 10^{-6}) and dispel separately in vials containing 5 ml sterile semi-solid N-free malate medium.

(3) Incubate the inoculated vials for 2 days at 32°C.

(4) Observe each vial for the presence of sub-surface pellicle of *Azospirillum* (which change the color of medium from green to blue).

(5) Transfer the loopful of growth from the subsurface pellicle developed in vial on the slants of NFB agar medium.

5.3   Cultural test to distinguish *Azospirillium* from contaminant

(1) On selective nitrogen-free malate semisolid medium in test tubes, the *Azospirillum* colonies form subsurface pellicle.

(2) On malate agar plate containing one percent NH_4Cl, the *Azospirillum* colonies are typically small and white dense (Fig. 6).

(3) On BMS agar (after 1–2 weeks of incubation at 33–35°C), the colonies of *Azospirillum* appears pink, opaque, irregular or round, often wrinkled and typical umbonate elevation (Tarrand et. al. 1978).

The colonies of *Azospirillum brasilense* on malate medium appears opaque and non-slimy, whereas the colonies of *Azotobacter chroococcum* on this medium are slightly viscous, semi-translucent during early growth and later turn dark brown (Tejera et. al. 2005).

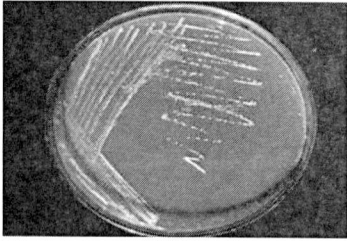

Figure 5.1 Azospirillium colonies on malate agar medium.

5.4   Test for Nitrogen fixation in pure culture of *Azospirillum*

The formation of white pellicle in semisolid Nitrogen-free Bromothymol blue media is the efficiency character of *Azospirillum* strain. Efficiency of Nitrogen fixation of *Azospirillum* is determined by kjeldahl Method.

5.4.1 Pure culture medium

Prepare Nitrogen-free Bromothymol blue medium for growth of pure culture of *Azospirillum*.

Procedure

Same as per the *Azotobacter*

Calculation

$$\text{Nitrogen content, percent by mass} = \frac{\text{Milliliters of 0.1 N acid for sample} - (\text{milliliters of 0.1 N acid for blank} \times 0.14)}{\text{Mass of sample taken}}$$

Total nitrogen in culture = Total dry mass of sample × percent nitrogen

6.1 Isolation of *Acetobacter diazotrophicus*

Different media viz., diluted cane juice semisolid medium, semisolid LGIP medium, acetified LGIP semisolid medium are all used for isolation of *Acetobacter*.

6.1.1 Materials

Sugarcane samples (root/leaf/stem/bud), 70% ethanol, Petri plates, water blanks, pipettes, LGIP semisolid medium, slants of these media, incubator, etc.

Diluted cane juice semisolid medium:

Semisolid LGIP medium	250 ml
Sugarcane cane juice	250 ml
Distilled water	500 ml

Semisolid LGIP medium:

Sucrose	100.0 g	K_2HPO_4	0.4 g
$MgSO_4 7H_2O$	0.2 g	KH_2PO_4	0.6 g
$FeCl_2$	0.01g	$CaCl_2$	0.02 g
Sodium molybdate	0.02 g	Agar	2.0 g
Distilled Water	1000 ml	Bromoethymol blue	5.0 ml
pH	5.5	(5% alcoholic solution)	

Semisolid Acetified LGIP medium:

Semisolid Acetified LGIP medium is acidified with acetic acid to pH 4.5 and agar concentration is increased to 2.2 g l^{-1}.

6.1.2 Procedure

(1) One gram of sugarcane samples (root/leaf/stem/bud) are washed thoroughly in running tap water and surface is sterilized with 70% ethanol, and subsequently washed in changes of sterile distilled water.

(2) The surface sterilized samples are macerated in a sterile blender.

(3) Prepare 10-fold dilutions of the above suspension up to 10^{-3}.

(4) One ml of 10^{-3} dilution is inoculated in to enrichment media.

(5) Enrichment culture is subculture for every 2–3 days.

(6) After enrichment, a loopful suspension is streaked on sterile petriplates containing LGIP medium. After that, incubate the plates at 28°C ($\pm$ 2°C) in an incubator.

6.1.3 Observations

(1) Observe for the development of smooth, flat with bright-yellow or yellow-orange color colonies of *Acetobacter*.

(2) Observe the dark-brown pigmentation of *Acetobacter* on potato dextrose agar medium (10% glucose).

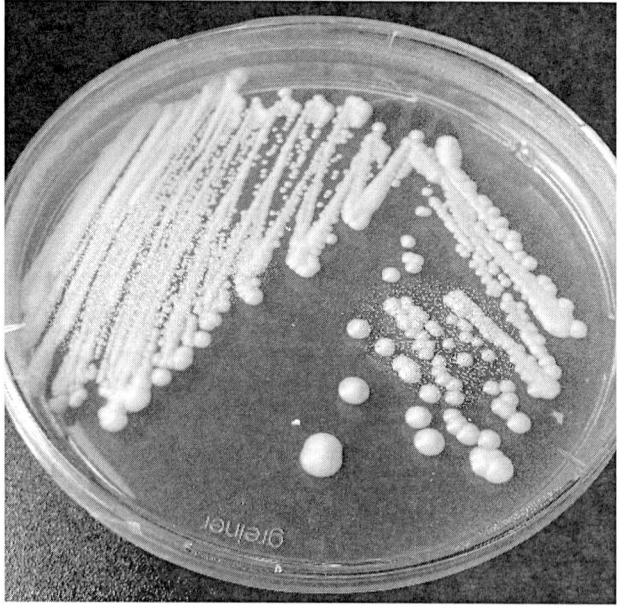

Figure 6.1 Colonies of *Acetobacter* on LGI medium.

6.2   Cultural tests to distinguish *Acetobacter diazotrophicus* from contaminatns

Acetobacter diazotrophicus is an acid lowing bacterium, requiring a pH optimum of 4.0–4.5 for growth and nitrogen fixation.

It requires high concentration of sucrose or glucose (100–300 g/l) for growth and nitrogen.

More importantly, this bacterium can fix nitrogen even in the presence of 25 mM of ammonia or 80mM of nitrate in medium. This property of this bacterium makes it distinctly different from other diazotrophus whose nitrogenase enzyme succumbs to even low level of combined nitrogen.

It is a gram-negative, non-spore forming, microaerophilic, diazotrophilic, endosymbiotic bacterium. It is a short and straight rod with round ends, measuring 0.7–0.9 × 0.2 μ. Cells are motile with either a single or three lateral flagella. Optimum growth temperature is 30°C and pH 4.0–4.5.

On LGI medium, the colonies of *Acetobacter diazotrophicus* appear irregular, 2–3 mm in diameter, smooth, flat with bright yellow or yellow-orange color (Fig. 7).

On PDA medium with 10% sucrose, the colonies would be dark brown; while on N poor agar (20 mg yeast extract per l) containing bromothymol blue, the colonies are dark orange.

Acetobacter diazotrophicus is more sensitive to antibiotics like rifampicin and tetracyclines and fairly resistant to streptomycin.

Cultures of *Acetobacter diazotrophicus* can be isolated without any difficulty from the root tissues of sugarcane or from the juice of crushed cane using nitrogen-free semi-solid LGIP medium.

Acetobacter and related species usually have a mucoid appearance when growing on agar and characteristic tan to brown colored translucent colonies.

Acetobacter species can use ethanol as carbon source, resulting in acetic acid production.

Bacterial growth occurs on up to 30% sucrose concentration but not at 35%.

By using selective growth media such as Hoyer's media, one can isolate *Acetobacter* from all other organisms. However, as some organisms do metabolize ethanol, the selective ingredient of Hoyer media, various tests should be performed on isolated colonies to determine whether or not they are *Acetobacter*.

6.3 Test for nitrogen fixation in pure culture of *Acetobacter diazotrophicus*

Nitrogen fixation efficiency of *Acetobacter diazotrophicus* is determined by kjeldahl method.

   (a) Pure culture medium:

Prepare Nitrogen-free LGIP medium for growth of pure culture of *Acetobacter diazotrophicus*.

   (b) Procedure

Same as per the *Azotobacter*

   (c) Calculation

$$\text{Nitrogen content, \% by mass} = \frac{\text{mm of 0.1 N acid for sample} - \text{mm of 0.1 N acid for blank} \times 0.14}{\text{Mass of sample taken}}$$

Total nitrogen in culture = Total dry mass of sample × percent nitrogen.

Blue-green algae are photoautotrophic free-living nitrogen-fixing organisms. Blue-green algae are generally covered with mucilage and it is easy to locate them as colonies floating on flooded rice fields. Blue-green algae being photoautotrophic use readily available energy and CO_2 carbon source to perform photosynthetic activities. N-free medium is to be used for isolation of blue-green algae from soil.

7.1 Isolation of blue-green algae from soil

Material required

Fogg's or Bristol's sodium nitrate liquid medium, conical flasks (250 ml capacity), water blanks, pipettes, illuminated growth cabinet, soil sample.

Fogg's medium

KH_2PO_4	0.2 g	$MgSO_47H_2O$	0.2 g
$CaCl_2$	0.1 g	Na_2MOO_4	0.1 mg
$MgCl_2$	0.1 mg	H_3BO_3 (Boric acid)	0.1 mg
$CuSO_4$	1.0 mg	$ZnSO_4$	0.1 mg
FE-EDTA	2.0 ml	Distilled Water	1000 ml

Procedure

(1) Prepare serial dilutions of soil or algal sample in sterile water blanks with sterilized pipettes.

(2) Prepare several conical flasks of nitrogen-free blue-green algae liquid medium (Fogg's medium).

(3) Inoculate aliquots of appropriate dilutions in to liquid media in flasks and incubate for several weeks in an illuminated growth cabinet at 28°C.

(4) As and when the individual colonies develop, they are transferred from the enrichment flasks to fresh aliquots of liquid media or on agar slant. Purification of culture is done by dilution method. Eradication of bacterial cultures with ultraviolet rays is recommended to avoid contaminants.

7.2   Cultural tests to distinguish blue-green algae from contaminants

Blue-greens are not true algae. They have no nucleus, the structure encloses the DNA and no chloroplast, the structure encloses the photosynthetic membrane. Blue-greens are more akin to bacteria which have similar biochemical and structural characteristics.

The majority of blue-greens are aerobic photoautotroph, their life processes require only oxygen, light and inorganic substances. Blue-greens can grow in full sunlight and in almost complete darkness.

The blue-green algae exhibit several different types of organizations. Some grow as single cell enclosed in a sheet of slime like material or mucilage. The cells of others aggregate into colonies that are either flattened, cubed, rounded, or elongated into filaments (Fig. 7.1).

The colonies look like tiny grey-green clumps (microcystis) or green, fingernail-like, or grass like clippings (Aphanizomenon).

When viewed under light microscope, blue-greens show a variety of movement such as gliding, rotation, oscillation, jerking, and flicking.

An oversupply of nutrients, especially phosphorus and possibly nitrogen often results in excessive growth of blue-greens to outcompete true algae. Blue-green adopt to the nutrition condition due to their positive buoyancy which is regulated by their gas vesicles which are absent in true algae.

Many blue-greens grow attached to the surface of rocks and stones, on submerged plants, or on bottom sediments of lakes (Fig. 7.2).

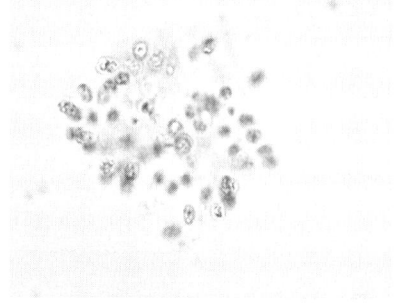

Figure 7.1 Flattened cells of blue-green algae.

Figure 7.2 Growth of blue-green algae on bottom sediments of lake.

Heterocysts and akinetes are unique to the blue-green. Blue-green can convert nitrogen gas into ammonia.

The iodine test is important to differentiate between blue-green algae and green algae. The green algae turn dark violet or black in color with iodine.

7.3 Test for nitrogen fixation in pure culture of blue-green algae

Nitrogen fixation efficiency of blue-green algae by Kjeldahl method:

(a) Pure culture medium:

Prepare Fogg's medium for growth of pure culture of blue-green algae

(b) Procedure:

Same as per the *Azotobacter*

(c) Calculation:

$$\text{Nitrogen content, \% by mass} = \frac{\text{ml of 0.1 N acid for sample } - \text{ ml of 0.1 acid for blank} \times 0.14}{\text{Mass of sample taken}}$$

Total nitrogen in culture = Total dry mass of sample × percent nitrogen.

8.1 Introduction

Azolla normally reproduce vegetatively by fragmentation of the abscission layer that forms at the base of each branch. Few species occasionally express gametophyte cycle and sporulated fronds are formed. The frond based spore inoculum of *A. microphylla* is used for multiplication.

Environmental factors are very much important in the multiplication and maintenance of *Azolla*. As *Azolla* is of aquatic habitat, an individual *Azolla* plant can survive only for a few hours on a dry surface under tropical conditions.

8.2 Isolation of *Azolla*

Fragmentation of the abscission layer of *Azolla* or *Azolla* frond is used for multiplication of *Azolla*. Under laboratory condition it is multiplied in small cement tanks (10 sq m). Following steps are to be followed in laboratory multiplication of *Azolla*:

(1) Maintain water level at 10 cm in tank/soil pond.

(2) Add 250 gm of fresh cow dung in tank water with 2.5 gm of super phosphate.

(3) Inoculate 200 gm of *Azolla* fronds in tank water.

(4) Apply furadan granules (2.5 gm) on the 7th day of fronds inoculation.

Similarly, frond based spore inoculums of *Azolla* (*A. microphylloa*) can be used. These frond based spore inoculums are pre-soaked in superphosphate solution (25 ppm) at a moisture saturation of 12 hours. Pre-soaked dried spores are released in 10 ml water and mixed well in tank water. Approximately, 1 gm of frond based spore inoculums is needed. The inoculated spores germinate well and sporelings emerge in 2 weeks period and multiply as *Azolla* (Fig. 8.1).

Figure 8.1 Growth of *Azolla* fronds.

The frond based spore inoculum of *A. microphylla* is used at 2–3 kg/ha for inoculation in transplanted rice field. The frond based spore inoculums are presoaked in superphosphate solution (25 ppm) at moisture saturation for 12 hours. Pre-soaked dried spores are released into 25 liters of water, mixed well and sprinkled in the main field uniformly for 7–10 days after transplanting. The inoculated spores germinate well and the sporelings emerge in 2 weeks period and multiply well in transplanted rice. *Azolla* biomass is effectively used as biofertilizer for rice. A layer of *Azolla* covering a hectare of rice field contains about 15–25 t biomass. The dry matter of *Azolla* usually contains 3–6% nitrogen. *Azolla* is decomposed in flooded rice field condition in 2–3 weeks period. *Azolla* can contribute 40–60 kg N/ha. The multiplication of *Azolla* along with rice crop also suppresses the aquatic weed populations. In addition to nitrogen release, *Azolla* also contribute potassium, phosphorus, calcium, sulphur, zinc and iron to rice soil. The dual culture method of growing *Azolla* with rice has perhaps the more widespread applicability because standing water is available in the field for growth of *Azolla* during the growth of rice from seedling to panicle initiation in most wetland rice fields. It grows harmoniously with rice plants and often remains green and healthy during such growth, being shaded from high light intensity by the rice canopy.

Fresh biomass of *Azolla* is broadcasted in the main field 7–10 days after transplanting the rice. Inoculation of the fresh biomass of *A. microphylla* at 200 kg/ha could multiply faster and could cover the rice fields as a green mat in 2–3 weeks period with 15–25 t biomass accumulation. *Azolla* technology is very efficient in terms of N_2 fixation and biomass accumulation during *rabi* season due to better environmental conditions (particularly, cloudy days coupled with low temperature favor its vegetative multiplication) prevailing during the second season rice. Growing *Azolla* along with rice crop is called dual culture and it does not affect the growth of rice crop in any way (Fig. 8.2).

Figure 8.2 Growth of *Azolla* in rice field.

For application of dried frond based spore inoculums of *Azolla*, 100 kg of sporulated fresh fronds of *A. microphylla* is formed into a heap. This is covered with a thin layer of clay soil slurry (20%) and allowed to undergo decomposition for a period of 21 days. Well-dried frond material is called "frond based spore inoculums" of *Azolla*. The viability of the spore inoculum is good for 10–15 months, which can be applied in rice field at 3 kg/ha.

8.3   Factors influencing the biomass production of *Azolla*

Environment influences the productivity of *Azolla* in rice production system. The primary constraint to the use of *Azolla* is its requirement for an aquatic habitat. An individual *Azolla plant* can survive only for a few hours on a dry surface under a tropical condition. It can survive for only a few days or a week on rice soil that dries during intermittent irrigations. Some varieties of *Azolla can* survive indefinitely on moist and shaded mud, but they will not multiply to any useful extent. For example, when enough irrigation water is available, relative growth rate, total biomass accumulation and nitrogen concentration are higher in cool dry environments. Environmental factors, such as temperature, light intensity and humidity are important in controlling the growth, multiplication and nitrogen fixation in *Azolla*.

8.3.1 Water

Azolla is very sensitive to dryness. It dies in a few hours if the soil becomes dry. *Azolla* growth is promoted by a fairly shallow depth of water (5 cm). Such a situation favors mineral nutrition since roots are near the soil and also reduces the negative effect on growth due to water turbulence. On the other hand, it should not allow rooting in the soil, which restricts growth, because it creates a tropic premature state of over population.

8.3.2 Temperature

The geographical distribution of *Azolla* clearly indicates that the genus is adapted to very different temperature conditions. Most of the *Azolla* species are widely distributed in temperate regions, as they are generally sensitive to the higher temperature of the tropics. Low tolerance to high temperature is one of the constraints to the multiplication of *Azolla* under field conditions. The optimum temperature range in which most of them grow well is 20–25°C. The temperature, particularly flood water temperature in rice fields, is a crucial factor, therefore the selection of strain for use in a given place at a given season must be done while taking temperature into special consideration. The influence of temperature on the growth of *Azolla* is dependent on many factors. Light intensity interacts with the effect of temperature. At higher light intensities, the optimum temperature shifts to higher temperature. The higher tolerance level of *A. microphylla* BR-GI and *Azolla* sp. – ST-SI to higher temperature is at 38 ± 1°C/25± 1°C day/night regime.

8.3.3 Humidity

The optimum relative humidity needed for normal *Azolla* growth is 85–91%. Low relative humidity below 60% causes *Azolla* fronds to dry up, turn fragile and become more susceptible to adverse conditions. High relative humidity causing a longer dew period results in susceptibility of the plant to diseases. In the tropics, high relative humidity during rainy season causes insect infestation.

8.3.4 Light intensity

Like other green plants, *Azolla* requires light for photosynthetic activity and the production of organic skeleton used in the cell synthesis reaction. In the tropics, direct sunlight during clear days at mid-day affects the growth and multiplication; while cloudy days provide a very favorable light exposure. Under high sunlight intensities, the fronds turn bricks-red which is a sign of physiological stress. Possibly to avoid maximum absorption of sunlight, the *Azolla* plant produces anthocyanin to protect its photosynthetic mechanism from damage. The growth and nitrogenase activity of *A. filculoides* changes with increasing light intensity. The growth increases with increasing light intensity to a maximum in 50% sunlight and CO_2 fixation saturated at 8000 lux in *A. carolliniana*. The growth rate of *Azolla* is higher at 25% sunlight (24500 lux) than full sunlight. The nitrogenase activity follows the same pattern, being highest in 50% sunlight and fractionally lower in 25% and 75% sunlight.

8.3.5 Wind

Wind tends to push all the fronds together on the same plant on the water surface. It is possible to reduce the influence of this factor by providing bunds and *Azolla*-rice intercrop.

8.3.6 Soil pH

The soil pH has a greater influence on the growth and multiplication of *Azolla* and the slightly acidic to neutral pH is found to be suitable for its growth. The very acidic soils of pH 3.0–3.8 do not support the growth. The soils having pH 5–7 support better growth than the soils of pH 8.0. *Azolla* can survive within a pH range of 3.5–10, but optimum growth is noticed at a pH range of 4.5–7.0. The growth of *Azolla* is reported to be good in nutrient solution at pH 5.5. The luxuriant growth of *Azolla* is produced at pH 5.5–6.6 in the irrigation water with soluble iron content of 0.56 mg/liter. It is also noticed that *Azolla* thrives at low pH in the presence of ferrous ions rather than that of ferric ions. An inverse relationship between pH and the temperature influences nitrate reduction and nitrogen fixation. Nitrogen reduction is optimum at pH 4.5 and 30°C, while nitrogen fixation is optimal at pH 6.0 and 20°C. The nitrogen fixation decreased at neutral pH.

Frankia: N$_2$-fixing endophytes

9.1 Isolation of *Frankia* from actinorhizal nodules

Frankia broth

Yeast extract	5.0 g	Dextrose	10.0 g
Casamino acid	5.0 g	Vitamin B	0.001 g
Distilled water	1000 ml	pH	6.4

(a) The nodules are thoroughly washed under running water to remove soil adhering to the surface. Most often, this treatment should be enhanced by carefully cleaning the nodule under a dissecting microscope to get rid of all soil and organic particles adhering to the nodule surface or inserted between the lobes.

(b) After fragmenting the nodule into individual lobes, nodule lobes are superficially sterilized by immersion in a 3% aqueous solution of osmium tetraoxide for 1–4 min according to the size and age of the nodule.

(c) Surface sterilized nodule lobes are then thoroughly rinsed with sterile distilled water and chopped into small pieces (0.1–0.5 mm^3) using a sterile scalpel. The treatment of nodule pieces with phosphate-buffered saline and polyvinylpyrrolidone (PVP-40) as recommended by Lalonde et. al. is optional.

(d) Nodule pieces are then evenly distributed onto a bottom layer of 1.5% agar nutrient medium in a petri dish. To increase the chances of success, the simultaneous use of several media: yeast extract-dextrose medium, Qmod medium, casamino-acids and sodium pyruvate medium or Qmod supplemented with activated charcoal are recommended. Media enriched with root lipids or with Tween 80 (polyxyethylene sorbitan mono oleate) are also recommended. Addition of cycloheximide at a concentration of 50 µg/ml may be useful in inhibiting growth of fungal contaminants without affecting *Frankia* growth.

(e) Pour about 3 ml of the same medium maintained at 40°C over the bottom layer, thus covering nodule pieces. Consequently, nodule pieces are embedded in the top layer, which provides microaerophilic conditions and facilitates the observation of developing *Frankia* colonies under a dissecting microscope.

(f) The petri dishes are sealed with parafilm and incubated at 28–30°C.

(g) After a 2–4 week incubation period, incubated petri dishes are observed for *Frankia* development using a dissecting microscope. *Frankia* colonies generally appear at the edge of nodule pieces.

9.2 Cultural test to differentiate *Frankia* from contaminats

- *Frankia* strains grow slowly, while contaminants during *frankia* isolations are fast growing.

- *Frankia* strains are difficult to isolate from soil and most strains originate from root nodules.

- *Frankia* has filamentous morphology (Fig. 9.1) and colonies are formed from spores or mycelia fragments after 7–10 days or longer on plating under best conditions.

- *Frankia* colonies are more prominent on defined propionate minimal medium of Baker and O'Keefe (Baker and O'Keefe 1984) or BAP medium (Murry et. al. 1984).

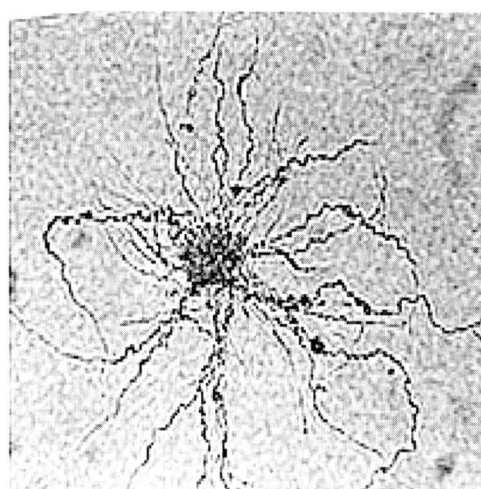

Figure 9.1 Filamentous *Frankia* colony.

9.3 Estimation of 'N' fixing efficiency of *Frankia*

Same as given in earlier chapters.

Phosphate solubilizing microorganism

10.1 Isolation of phosphate solubilizing microorganism

Fixation of phosphorus in the soil which is supplied through chemical fertilizers, poses the problem of availability of this essential element to crops; and therefore, crops suffer from P deficiencies. Some of the microorganisms present in soil solubilize this fixed form of phosphate and make it available to the crops.

Phosphate solubilizing microorganisms can be isolated from soil by using Pikovskaya's medium containing tricalcium phosphate. On plating aliquot of soil dilution, the P solubilizing microorganisms show clear zones of P solubilization around the microbial colony on this medium. Such colonies are picked up and sub cultured; and thus, pure culture of P solubilizing microorganisms can be maintained on slants of Pikovskaya's medium.

10.1.1 Materials

- Rhizosphere soil of legume crop
- 9 ml sterile water blanks, sterile petri plates
- Pipettes
- Inoculating needle
- Incubator
- Test tubes
- Pikovskaya's medium

10.1.2 Pikovskaya's medium

Glucose	10.0 g	$Ca_2(PO_4)_2$	5.0 g
$MgSO_4\ 7H_2O$	0.1 g	KCl	0.2 g
$(NH_4)_2\ SO_4$	0.5 g	$MnSO_4$	Trace

| FeSO$_4$ 7H$_2$0 | Trace | Yeast extract | 0.5 g |
| Agar | 15.0 g | Distilled water | 1000 ml |

10.1.3 Procedure

(1) Prepare 10-fold dilutions serially up to 10^{-6}.

(2) Transfer 1 ml each of 10^{-2}, 10^{-3}, 10^{-4} and 10^{-5} dilution aliquots in quadruplicate plates separately.

(3) Pour about 15 ml molten and cooled Pikovskaya's medium into the above plates and mix the soil suspension by gently rotating the plates and allow the medium to solidify.

(4) Incubate the plates at 30°C for 3–4 days.

10.1.4 Observation

(1) Observe for the growth of microbial colonies showing clear zone around them (Fig. 13).

(2) Count such a colony and calculate the number of P solubilizing microbes in original soil samples by following formula:

No. of P solubilizers/g soil = Av. no. of colonies × Dilution factor with clear zones

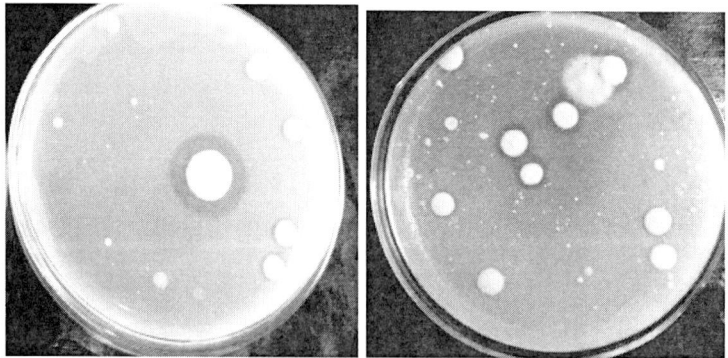

Figure 10.1 Colonies of phosphate solubilizing bacteria showing clear zone around them.

10.2 Estimation of phosphate solubilizing ability of phospho solubilizer

The ability of the bacterial isolates to solubilize insoluble inorganic phosphate is tested by spotting 10 µl overnight cultures on Pikovskaya's agar plates and incubating at 28–30°C for 2–3 days. The isolates which show clear zone of solubilization of tricalcium phosphate (TCP) around the colony are noted as phosphate solubilizers. The diameter of the zone of TCP solubilization is measured and expressed in millimeters.

10.3 Quantification of soluble phosphorus released by phosphosolubilizer

10.3.1 By using ammonium molybdate reagent

Reagent

(1) *Chlorostannous acid*: 2.5 gm of $SnCl_2.2H_2O$ dissolved in 10 ml of concentrated HCl (heat). Make volume to 100ml with distilled water.

(2) *Chloromolybdic acid*: 15.0 gm of ammonium molybdate dissolved in 400 ml of warm distilled water. Add 342 ml of 12 N HCl and cool it down. Make the volume up to one liter with distilled water.

Procedure/Quantitative test

(1) Prepare Pikovskaya broth with known amount of inert phosphorus (Rock phosphate/Tricalcium phosphate).

(2) Inoculate it with test microorganisms.

(3) Incubate at 28–30°C on shaker for 3–4 days.

(4) In case of bacterium, the broth is centrifuged at 10,000 rpm for 10 minutes; and it is filtered in case of fungal inoculation.

(5) Take aliquot (0.1–1.0 ml) from the supernatant/filtrate and add 10ml of reagent II (ammonium – molybdate). Shake well. Dilute the contents of the flask to 45 ml by adding distilled water. Add 0.25 ml of reagent I (chlorostannous acid) and immediately make up the volume to 50 ml.

(6) Measure the blue colored intensity of the solution at 600 nm in spectrophotometer.

(7) Find the corresponding amount of soluble phosphorous from the standard curve.

Precautions

• Make the reagent chlorostannous acid just before use.

• Store ammonium molybdate in amber colored bottle.

10.3.2 By using ascorbic acid reagents

Reagent

(1) Ammonium molybdate

(2) L-Ascorbic acid

(3) p-Nitrophenol indicator

(4) 4N H_2SO_4

Sulphomolybdic acid

(1) Take 20 gm of ammonium molybdate and dissolve it in 300 ml of distilled water.

(2) Slowly add 450 ml of 10 N H_2SO_4.

(3) Cool down the above mixture. Add 100 ml of 0.5% solution of antimony potassium tratrate.

(4) Cool and make the volume to one liter in glass bottle and store away from direct sunlight.

Preparation of mixed reagent I

Add 1.5 g of L-ascorbic acid in 100 ml of the above stock solution and mix it. Add 5 ml of this solution to developed color. Mixed reagent is to be prepared fresh as it does not stay for more than 24 hours.

Procedure/Quantitive test

(1) Take a known aliquot (5–25 ml) of the extract as mentioned in previous procedure in a 50 ml volumetric flask.

(2) Add 5 drops of p-nitro phenol indicator (1.5% solution in water) and adjust the pH of extract between 2 and 3 with the help of 4N H_2SO_4. The yellow color disappears when the pH of solution becomes 3. Swirl gently to avoid loss of the solution along with the evolution of CO_2.

(3) When the CO_2 evolution has subsided wash down the neck of the flask and dilute the solution to 40 ml.

(4) Add 5 ml sulphomolybdic acid mixed reagent containing ascorbic acid, swirl the content and volume.

(5) Measure the transmission after 30 min at 880 nm using red filter. The blue color developed remains stable up to 60 minutes.

(6) Record concentration of phosphorus (P) in the extract from the standard curve.

Preparation of standard Curve

Prepare standard curve using 0.1–0.6 ppm P in 50 ml volumetric flask. Plot the curve by taking concentration of soluble p on x-axis and percent T on y-axis using a semi-log graph paper. It is straight line relationship between the soluble p and percent T on a semi-log graph paper.

Vesicular arbuscular mycorrhizae

11.1   Introduction

The first step in a mycorrhiza research programme is the isolation of VAM spores from soil, so that they can be multiplied, screened for effectiveness and used subsequently as inoculum. To find VAM fungi effective on a nitrogen-fixing tree species, it is advisable to collect soil from an area where the occurrence of the host species is common. At the same time, isolates from other soils should be obtained to test for effectiveness.

After removing debris from the soil surface, soil samples should be collected to a depth of 15 centimeters. Before recovering VAM spores, they can be increased by growing a 'trap plant' in the soil (Hung et. al. 1985). This procedure is particularly important for isolation of VAM species which do not sporulate abundantly. Wet sieving and decanting (Gerdemann 1955; Gerdemann and Nicolson 1963) are the most commonly used techniques, and they enable the separation of spores on the basis of size. Further purification of spores is done by sucrose centrifugation. Techniques are available for obtaining single-spore cultures from soil (Francis and Reid 1984). After the spores have been separated from the soil, it is important to characterize them and if possible, they should be identified. Observation of the spores with regard to mode of development, morphology of hyphal attachment, size, shape and color and finally, wall structure is to be recorded.

11.2   Isolation of arbuscular mycorrhizal fungal spores

- Isolate the arbuscular mycorrhizal fungal spores from soil samples collected from root rhizophere of mango tree.

- Collect the samples by trowel to a depth of 10–15 cm by scrapping away the top 1–2 cm layer of soil.

- 50 g soil of 10 mango trees is taken and mixed thoroughly form which 50 g is taken for isolation of AM fungal spores.

- Isolation of AM fungal spores is carried out by wet sieving and decanting technique.

- 50 gm of soil sample is suspended in luke warm water in a large beaker, until all soil aggregates have dispersals to leave a uniform suspension.

- Most of the suspension is decanted through 710 µm sieve and collected into 1 lit graduated cylinder.

- The residues are re-suspended in more water and decanted.

- This process is repeated 4–5 times to get about 700 ml in a cylinder with five sets of water.

- The roots and other organic matter on the sieve are washed and the washing is collected in the cylinder. The material in cylinder is re-suspended by stirring several times and decanted through 250 µm sieve into a second cylinder, retaining a small volume which is re-suspended in 300 ml of water and poured through the sieve to fill the second cylinder.

- The material from 2nd cylinder was re-suspended in another 1 lit water and poured through 53 µm sieve.

- Each sieving was washed into a separate small beaker and transferred into the Doncaster nematode counting dish and examined under stereoscopic zooming microscope. Spore count after each harvest is also taken by this method.

11.3   Identification of VAM fungi

11.3.1 Spore morphology of VAM fungi

(1)   *Glomus epigaeum*:

Spore shape: Globose or subglobuse, ovate or irregular.

Spore size: 150–175 µm, spores occurred in cluster.

Nature: Smooth, dark yellow to brown, hyaline, spores are smaller than Glomus mosseae.

Colour: Reddish-brown to orange-brown, outer layer hyaline, inner yellow to brown.

Stalk attachment: Cylindrical or flared towards the point of attachment or constricted one.

Wall number and thickness: Double wall layer.

(2)   *Glomus mosseae:*

Spore shape: Globose, often mixed with sub-globose, ovate, ellipsoid, or irregular shapes.

Spore size: 100–300 μm in diameter, occurred singly, large spore.

Nature: Smooth spores formed ectocarpically or within sporocarps of 1–10 spores.

Color: Spores are yellowish-brown to brown, outer layer hyaline, inner layer is yellowish to brown.

Stalk attachment: Simple, straight, recuned or funnel shaped hyphal attachment.

Wall number and thickness: Two wall layers; outer is thin and inner is thick.

(3)   *Gigaspora calospora*:

Spore shape: Globose, ellipsoid, obovoid, reniform, or irregular shape.

Spore size: 250 μm or larger.

Nature: Smooth to dull, roughened with slender spines borne singly.

Color: Hyaline, white or shades of yellow.

Stalk attachment: Bulbous, suspensor like cell, usually with one to several emergent pegs, spores produced on tip or side of cell.

Wall number and thickness: Single layered, usually less than 5 μm thick.

11.4   Inoculum production

Once VAM spores have been isolated from the soil, they have to be increased before they can be screened for effectiveness and used as an inoculum. The strict symbiotic nature of VAM fungi necessitates that they should be grown in the presence of a plant root (Fig. 11.1). This limitation has been a major constraint in the mass production of VAM fungi. Jarstfer and Sylvia (1992) have updated the techniques used in VAM inoculum production. In brief, production of VAM inoculum involves surface disinfestations of spores which are then placed near the roots of young seedlings. The seedlings are grown for approximately three months with sanitation to avoid contamination. The cultures thus produced can be stored in a dry condition for a considerable period of time without loss of viability. The parent cultures should be checked periodically for purity and viability. Maintenance of good germplasm should be an essential part of a VAM research programme. The International Culture Collection of VAM Fungi (INVAM) maintains a collection of VAM.

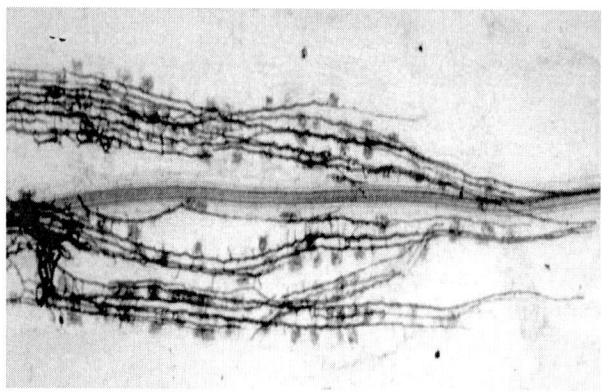

Figure 11.1 Spores of VAM fungi on plant root.

More recently, VAM inoculum has been produced in hydroponic and aeroponic cultures (Hung and Sylvia 1988). In a hydroponic system, plant roots are bathed in a circulating nutrient solution; whereas in an aeroponic system, roots are grown in a fine continuous mist of nutrient solution. The aeroponic system produces inoculum that is cleaner and of higher quality than those produced in a soil-based system. Sylvia and Jarstfer (1990) introduced the concept of sheared-root inoculum. With this technique, the colonized roots produced in an aeroponic system are sheared with a food processor to form a slurry. The slurry can be incorporated with a carrier or encapsulated in the form of pellets. Based on the cost-benefit analysis of this product, sheared-root inoculum from aeroponic cultures should give an economical basis for practical utilization of VAM fungi under some conditions. Regardless of the technique used to produce an inoculum, the product should be pure, effective, and at the same time economical.

11.5   Mycorrhizal dependency

Mycorrhizal dependency is the degree to which a plant is dependent on the mycorrhizal condition at a given level of soil fertility. It is estimated as percentage by the formula given by Plenchette et al. (1982) as given below:

$$\text{Mycorrhizal dependency (\%)} = \frac{\text{DM of inoculated plants} - \text{DM of uninoculated plants}}{\text{DM of inoculated plants}} \times 100$$

Where, DM = Dry matter weight.

Depending on the mycorrhizal dependency percentage, the plant is said to be highly dependent (75% and above), dependent (50–74%), moderately dependent (25–49%), marginally dependent (up to 24%) and independent when not colonized by VAM (0% mycorrhizal dependency).

11.6 Screening for effective isolates

Since isolates of VAM fungi vary dramatically in their effect on plant growth, it is necessary to screen them for their effectiveness. To screen VAM fungi, isolates are inoculated into a host and the plants are grown in soil under uniform greenhouse conditions (Hung et. al. 1990). The isolates that produce maximum host growth are selected. Alternatively, a non-destructive technique that involves monitoring phosphorus status of sub-leaflet or leaf disc of indicator plants could be utilized. It is important that the selected VAM fungi are able to compete with native VAM fungi. To obtain such an isolate, the fungi should also be tested in the field where the introduced VAM fungus will compete with the native ones. When VAM isolates are to be compared for effectiveness, the inoculum potential of the fungi should be equalized. This can be done by the Most Probable Number (MPN) technique (Plenchette et. al. 1982; Powell 1980). However, Sylvia and Burks (1988) have raised concern about the reliability of the MPN test for determining the VAM inoculum potential. Caution should therefore be exercised in interpreting MPN results for VAM effectiveness. This assay can also be used to assess the infectivity of VAM inoculum when it is ready for use.

11.7 Application of inoculum

To increase the chance of contact between roots and fungus, the inoculum can be placed in bands or layers just below the seeds. Plants can also be inoculated with selected isolates of VAM fungi at the seedling stage in a nursery and then transplanted to the desired site. Such a procedure would be particularly useful in reforestation or soil rehabilitation programmes. Application of inoculum in the form of pellets is a relatively new technology; nonetheless, pelleting of seeds with VAM inoculum along with a carrier substrate has shown promise in certain crops. The advantage of this method is that *rhizobia* can also be introduced to nitrogen-fixing trees along with VAM fungi.

11.8 Arbuscular mycorrhizal fungal colonization

It is necessary to identify arbuscular mycorrhizal fungal infection in pot culture and in field. The anatomical features characteristics of arbuscular mycorhizal fungi cannot be seen unless the infected roots suitably stained. A commonly used staining technique, i.e. Phillips and Hayman (1970) is used to identify arbuscular mycorrhizal fungal infection, which is described as follows:

(1) The roots are thoroughly washed in tap water and cut into one centimeter length.

(2) They are then immersed at 90°C for 1 hour in 10% KOH solution.

(3) The KOH solution is decanted and root bits are rinsed four times in tap water.

(4) The root bits are acidified in dilute 2% HCL for five minutes.

(5) The acid is poured and the root bits are stained with 0.05% tryptophan blue boiled for 3 minutes.

(6) The excess stain poured off and lactophenol (Lactic acid + Glycerol) is added, which is allowed to stand overnight to distain the host tissue.

The root segments are then mounted on slides in lactophenol. Slight pressure is given on the cover slip to flatten the roots for observation within a limited range of focus under compound microscope. Twenty five root segments are collected at random for each experimental treatment and the root percent colonization is calculated as follows:

$$\text{Am fungal coonization (\%)} = \frac{\text{Number of AM positive segments}}{\text{Total number of root segments observed}} \times 100$$

11.9   Estimation of infectious propagule per gram of soil

Inoculum potential is expressed as the number of propagule per gram of soil. The procedure is similar to that described by Plenchette et. al. (1989) with some modifications. The inoculated soil samples with trap crop roots after each harvest/at final harvest, as the case may be, are collected. Eight seeds of *Zea mays* are planted per pot with 1 gm of AM soil sample and cultivated for 12 days in a pot culture (30 ± 2°C, RH 60%). Roots are washed and stained by standard procedure of Phillips and Hayman (1970) and are observed for estimation of primary infection.

The number of primary entry points is counted on whole root system under a stereoscopic zooming microscope. Infectious propagule density of AM is calculated on the basis of primary entry points on the root system.

Sulphur oxidizing microorganism

12.1 Isolations of sulphur oxidizing bacteria

Collect the sulphur enriched rhizosphere soil from a depth of 22.5 cm or soil from sulphur hot-spring and air-dry under shed. The stones, pebbles, etc., should be separated and the soil should be grinded in a wooden pestle and screened through 2 mm sieve. Isolation is carried out by an enrichment techniques using sulphur enriched medium. The pH of the medium is adjusted to 5 and 8.0. Five ml of bromocresol purple (0.25%) indicator is added to per litre medium and the medium is then distributed in each of the 250 ml Erlenmeyer flasks. The flasks are sterilized in autoclaved at 1 kg/cm^2 pressure for 20 min for three successive days. The flasks are then inoculated with one gram of soil and incubate on to-and-fro shaker (150 to-and-fro actions) at room temperature for eight day. The flasks showing the fall in the pH of the medium is evidenced by the color change of the bromocresol purple indicator from purple to colorless. Such flasks are presumed to contain the growth of sulphur oxidizing bacteria.

12.2 Purification of sulphur oxidizing bacteria from enriched medium

For purification of sulphur oxidizing bacteria, thiosulphate medium containing sodium thiosulphate is used. The pH of the medium is adjusted to 5 and 8.0 by using 0.1 N NaoH or 0.1N Hcl as required. Five milliliter of bromocresol purple (0.25%) indicator is added to a litre of medium before sterilization. A loopful of the inoculum from the enrichment broth showing positive indication for the presence of the sulphur oxidizing bacteria is streaked on the thiosulphate medium in duplicate plates. The medium of the different pH values is to be tried to make out the isolates with different pH optima. The plates are then incubated for 5 days at 30°C. The colonies producing the acid, as indicated by the change in the indicator from purple to colorless are sulphur oxidizing bacterial colonies (Fig. 12.1). Such colonies are selected and streaked on to the slants of the fresh thiosulphate agar medium of the same composition and pH.

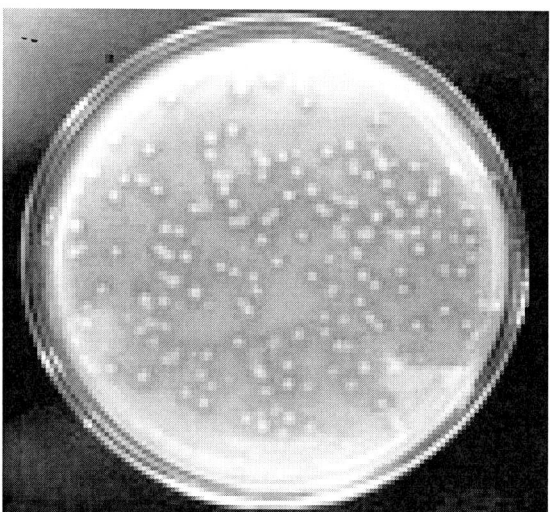

Figure 12.1 Bacterial colonies of sulphur oxidizer.

12.3   Isolations, purification and maintenance of sulphur oxidizing fungi and actinomycetes

For isolation purpose, an enrichment technique is used. The medium consists of sodium thiosulphate as sulphur source and sucrose as carbon source. The pH is adjusted to 5.0 and 8.0 by using 0.1 N NaoH and as required bromocresol purple (0.25%) indicator is added for detection of the colonies. Presumptive sulphur oxidation is indicated by the formation of zone of clearing around the colony of fungi and actinomycetes.

12.4   Screening of isolates for sulphur oxidizing efficacy

The bacterial isolates are inoculated in the sulphur liquid medium and incubated for 20 days. Similarly, fungi and actnomycetes are inoculated in modified sulphur liquid medium. The sulphur oxidizing efficacy of the different isolates are judged by the reduction of pH of liquid medium (Fig. 12.2) and sulphate formation from the oxidation of elemental slulphur. The pH of the medium is measured by digital pH meter and sulphate sulphur is determined by turbidimeterically by using spectrophotometer (spectronic 20) at 490 nm.

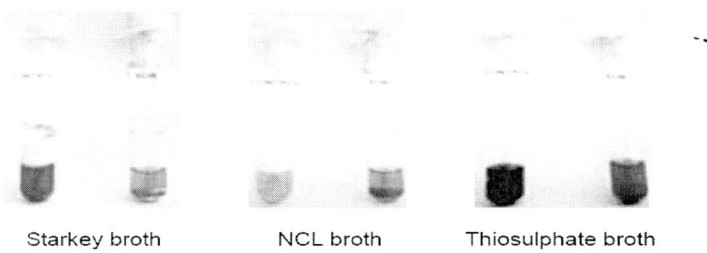

Starkey broth   NCL broth   Thiosulphate broth

Figure 12.2   Reduction of pH in the growth media by sulphur oxidizing
bacteria.

Silicate solubilizing biofertilizers

13.1 Introduction

Silicate solubilizing bacteria (SSB) release silica (SiO_2) in solution from the water-insoluble silicates. This principle is employed in isolating SSB from soil and other materials. SSB is distributed in cultivated soils, agro-inputs rock and super phosphate, cement pipes, mountain rocks and sand, which is purely quartz and soil in granite crusher yard.

13.2 Isolation and enumeration

SSB is isolated from soil samples after serially diluting samples up to 10^{-3} dilution. One ml aliquot of appropriate dilution is plated in Modified Bunt Rovira medium containing 0.25% insoluble magnesium trisilicate and the plates are incubated at room temperature ($30 \pm 2°C$). The colonies exhibiting clear zone of silicate solubilization are SSB (Fig. 13.1). Such colonies are counted and expressed number per gram of dry sample.

13.2.1 Modified bunt and rovira medium

Peptone	1.0 g	Yeast extract	1.0 g
Glucose	20.0 g	Ammonium sulphate	0.50 g
Magnesium chloride	0.10 g	Dipotassium hydrogen phosphate	0.40 g
Ferric chloride	0.01 g	Magnesium trisilicate	2.50 g
Soil extracts	250.0 ml	water	750.0 ml
Agar-agar	20.0 g	pH	6.6–7.0

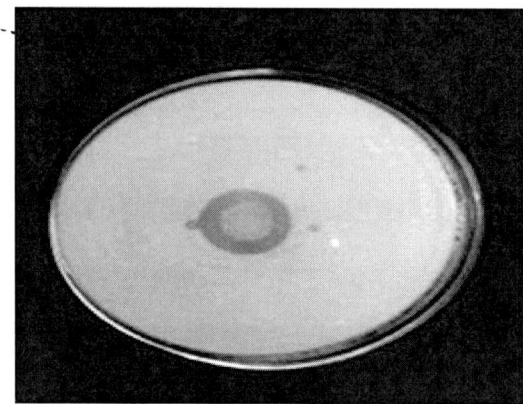

Figure 13.1 Silicate solubilzing bacteria showing clear zone on Bunt and Rovira medium.

13.3 Testing of silicate solubilization potential

The bacterial isolates exhibiting clear zones in the agar plate containing medium with insoluble silicate is picked up and maintained in nutrient agar slants.

These cultures are inoculated to 50.0 ml of a basal medium containing 0.25% magnesium trisilicate taken in 100.00 ml Erlenmeyer flask and the release of silica is estimated by the method described by Saxena (1989).

13.3.1 Basal medium

Glucose	10.0 g	$(NH_4)2 SO_4$	1.0 g
KCl	0.2 g	K_2HPO_4	0.1 g
$MgSO_4$	0.2 g	Distilled water	1000 ml
Magnesium trisilicate	2.5 g	pH	7.0

13.4 Estimation of available silica (Saxena 1989)

13.4.1 Reagents

(1) *Reagent A*: 50 ml of conc. HCl is mixed with 50 ml of distilled water.

(2) *Reagent B*: Ammonium molybdate 10 g is dissolved in 100 ml of distilled water, the pH is adjusted between 7 and 8 with ammonium hydroxide and stored in polyethylene bottle.

(3) *Reagent C*: Oxalic acid 10 g is dissolved in 100 ml of distilled water.

13.4.2 Procedure

Aliquot of 5/50 ml is taken in polyethylene beaker to which 1 ml of Reagent A and 2 ml of Reagent B are added. After 10 min, 1.5 ml of reagent C is added and mixed thoroughly. The yellow color developed is read at 410 nm in a Baush and Lomb spectronic 20 colorimeter. The unknowns are calculated from a standard curve prepared by dissolving 0.2 g of sodium metasilicate in 100 ml of distilled water (0.566 mg/1 SiO_2) which is further diluted to series of standard from 1–10 mg silica (SiO_2) per 50 ml.

Potassium solubilizing biofertilizers

14.1 Isolation of potassium solubilizing microorganisms

- Potassium solubilizing microorganisms are isolated from soil samples by serial dilution (10^{-3} for bacteria, 10^{-2} for fungi and actinomycetes) plate count method on wring Alesksandrov medium which is a selective medium for isolation of potassium solubilizers.

- Incubate the plates at room temp ($20 \pm 1°C$) for 3 days and select the colonies exhibiting clear zones and purify by four way streak plate method.

14.1.1 Media composition

Aleksandrov media (pH – 7.2)

(1) Glucose – 5.0 g

(2) Magnesium sulphate – 0.005 g

(3) Fecl$_3$ – 0.1 g

(4) Calcium carbonate – 2.0 g

(5) Potassium mineral – 2.0 g

(6) Calcium phosphate – 2.0 g

(7) Agar – 18.0 g

(8) Distilled water – 1000 ml

14.2 Testing of effective strains of KSM

Measure the diameter of zone of solubilization and express it in centimeter (Fig. 14.1).

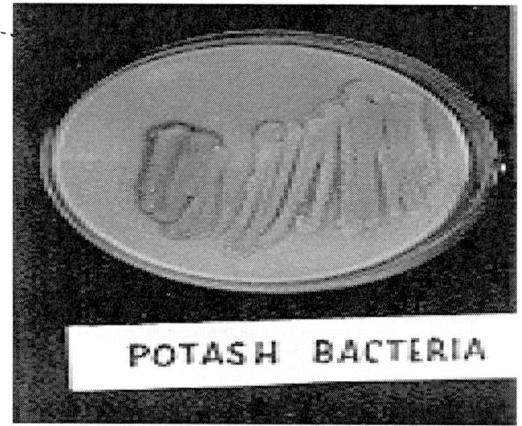

Figure 14.1 Zone of solubilization produced by KSM on Aleksandrov media.

14.3 Quantitative estimation of k released from insoluble k

(1) Inoculate 1 ml of overnight culture of KSM in 25 ml of Aleksandrov broth.

(2) Incubate the inoculated flasks for one week at $28 \pm 2°C$.

(3) Centriguge the broth culture at 10,000 rpm for 10 min in the semi-micro centrifuge to separate supernatant from the cell growth and insoluble potassium.

(4) Determine the available k content in the supernatant by flame photometry.

(5) Estimate the efficacy of potassium solubilizing microorganisms from the standard curve prepared for potassium.

14.3.1 Preparation of standard curve

1.908 g of kcl is dissolved in 1 litre distilled water. Dilute 10 ml of this to 100 ml with distilled water to obtain 2 ppm solution and use it for preparation of standard 0, 2, 4, 6, 8, 10 ppm. Analyze these standards in flame photometer to obtain k standard curve.

15.1 Isolation and maintenance of decomposting cultures

15.1.1 Materials

Soil sample, sterile water, test tubes (water blanks), petri plates, cellulose agar medium

Cellulolytic media for fungal isolation (Asparagines medium)

$(NH_4)_2 SO_4$	0.5 g	L-Asparagine	0.5 g
$MgSO_4 7H_2O$	0.2 g	$CaCl_2$	0.1 g
KH_2PO_4	1.0 g	KCl	0.5 g
Cellulose	10.0 g	Yeast extract	0.5 g
Distilled Water	1000 ml	pH	6.2
		Agar	15 g

Cellulolytic media for Bacterial isolation (Han's medium)

$(NH_4)_2 SO_4$	1.0 g	$MgSO_4 7H_2O$	0.1 g
$CaCl_2$	0.1 g	KH_2PO_4	0.5 g
NaCl	6.0 g	K_2HPO_4	0.5 g
Cellulose	10.0 g	Yeast Extract	0.1 g
Distilled Water	1000 ml	pH	6.5
		Agar	15 g

15.1.2 Procedure

(1) Add 1 g soil sample to 9 ml sterile water in test tube. It will give 10^{-1} dilution.

(2) Transfer 1 ml suspension of 10^{-1} dilution to next 9 ml sterile water blank and mix thoroughly by shaking for 1 minute. This will give 10^{-2} dilution of original soil sample.

(3) Repeat the above steps and prepare 10^{-3}, 10^{-4} 10^{-5}, 10^{-6}, 10^{-7} and 10^{-8} dilutions of the soil sample in test tubes containing sterile water.

(4) Transfer 1 ml suspension from each dilution into sterile petri plates separately.

(5) Pour the plates with molten cellulose agar medium. Mix the soil dilution aliquot medium by gently rotating and allowing the medium in plates to solidify.

(6) Incubate the plates at 30°C for 8–10 days.

(7) Count the number of colonies of microorganisms showing a change in color around the colonies (from grass-green to yellow) which is an evidence for the production of acid from degradation of the cellulose. Pick up these colonies of cellulolytic microorganisms and transfer them to the slants of cellulose agar medium or potato dextrose agar medium. Maintain the cultures by periodic transfers to the fresh slants of these nutrient media within six months.

15.1.3 Observations

(1) Count the number of cellulolytic microbial colonies in each plate and compute the average number per gram of soil.

(2) The number of cellulolytic microorganisms of oven-dry soil is calculated by multiplying the average plate counts by the dilution factor and dividing by the oven-dry weight of one gram of soil.

$$\text{No. of cellulolytic microbes or oven dry soil} = \frac{\text{Average plate count} \times \text{Dilution factor}}{\text{Oven dry weight of 1 g of soil sample}}$$

(3) Record the colony morphology of cellulolytic microorganisms (Fig. 15.1).

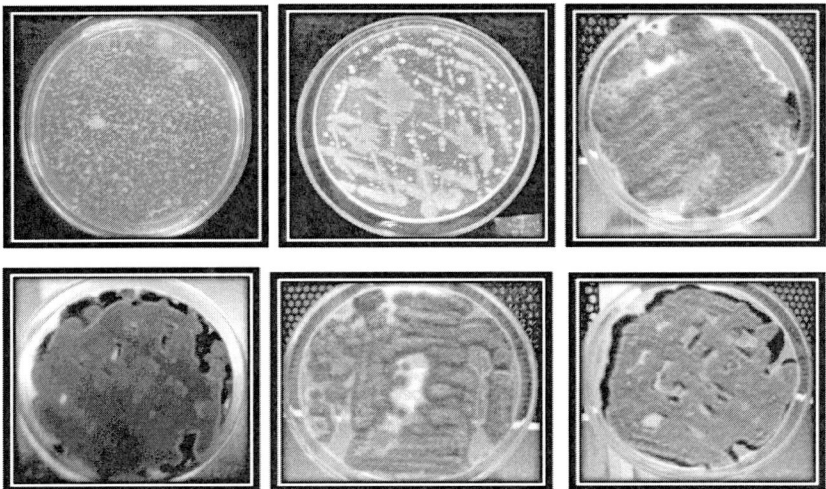

Figure 15 Decomposing microorganisms.

Production technology of biofertilizers

16.1 Design for biofertilizer production unit

For production of biofertilizers, a well-equipped laboratory is required. It is necessary to have separate rooms for inoculation, fermentation, mixing and curing, storage and packing of biofertilizers. There must be sufficient space for movement from one room to another. The design, layout and equipment selection must be thoughtfully made for each production facility to meet their specific needs.

16.1.1 Laboratory

The laboratory should be designed for bacteriological work. There should be adequate facilities and equipments for quality control of the cultures and inoculants (Fig. 16.1). Laboratory bench-space, gas outlets, sinks, laboratory equipments, large autoclave and storage facilities are essential components. It is necessary to provide an area in which the transferring of cultures between tubes and flasks and dilution plating can be done with a minimum chance of contamination. The need for making aseptic bacterial transfers is very important in the maintenance of a culture collection, as well as during the transferring for culture production. A laminar-flow sterile air-bench in a separate transfer room provides excellent protection for aseptic bacteriological transfers. The transfer room should have ultraviolet lights, gas outlets for laboratory burners, and electrical outlets.

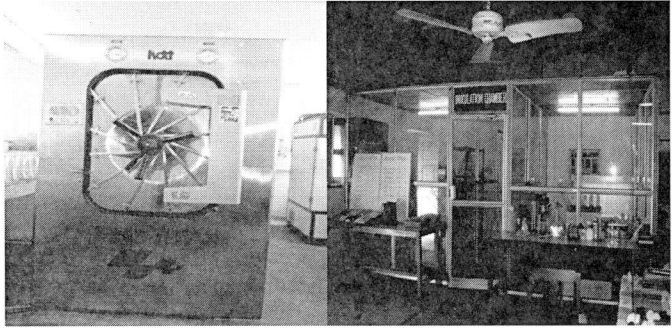

Figure 16.1 Microbial culture isolation laboratory.

16.1.2 Fermentor room

The fermentor room should be designed for large scale culture production and must maintain a high degree of sanitation. It is advisable at the time of initial plant installation to locate equipment so that all plumbing and controls have convenient access; and to allow space for additional equipment at a later time when increased production needs warrant.

16.1.3 Mixing and curing room

A clean environment is essential in the room used for aseptic culture mixing and curing in sterile production system. The location of the room utilized for mixing of the culture carrier should be adjacent to the fermentor room. However, if the dust is generated in the mixing process, it is essential not to have direct air passage or doors connecting the fermentor room with the mixing room. The mixed inoculant may be cured in a suitable mixing room. Temperature and humidity in the curing area should be maintained to provide optimum culture growth in the carrier without appreciable moisture loss.

16.1.4 Packaging room

The packaging area should be next to the curing room. Supplies required for packaging inoculant, such as polyethylene bags, cartons and labeling materials should be readily accessible to the packaging area. The packaging operation, whether hand filled or automated is most efficient in an assembly line operation. Sufficient space in the room for filling the bags, heat-sealing the bags and assembling into a carton is required.

For packaging, flexible low-density polyethylene bags (0.038–0.076 mm) are used. These polyethylene bags provide high moisture retention, sufficiently high gas exchange and is heat sealable. The task of filling the inoculant bag with the cured inoculant can be labor intensive with hand-filling or a semiautomatic weight dumping machine. Both processes require personnel filling of bags, checking for an appropriate weight, placing pinholes in the bags for aeration, and then sealing the top of each bag. This complete process can be automated with a bag form, fill and seal machine. A continuous roll of preprinted film is automatically formed into a bag, volumetrically filled, pin-holed and heat-sealed with great speed. However, large volumes must be produced to warrant investment in an automatic system.

The package should provide the proper product identification specifying the strains for which it may be used; detailed application instructions for the user, an expiry date, quantity of seeds, or area to be treated and the manufacturing company.

16.1.5 Storage room

Adjacent to the packaging room should be the finished goods storage room, providing temporary storage of the product. In temperate climate, ambient temperatures and humidity may be utilized; however, air conditioner is advisable and refrigerated storage is ideal in warm weather. Bioinoculants are perishable products which should not be exposed to enhance drying such as increased temperature and direct sunlight.

16.2 Production technology

16.2.1 Materials

Nutrient liquid media required by specific N_2 fixing organisms, i.e. Jensen's liquid medium for *Azotobacter*, yeast extract mannitol broth for *Rhizobium* and nitrogen-free broth for *Azospirillum*, efficient cultures of diazotrophs, suitable carrier material, e.g., peat, lignite, etc., autoclave, shaker, polythene bags, flasks, inoculating needle, sealing machine, etc.

16.2.2 Procedure

Production of biofertllizers involves the following steps:

(1) Preparation of starter culture

(2) Preparation of broth culture

(3) Preparation of carriers

(4) Preparation of inoculants in powder form

   (a) *Preparation of starter culture*: Pure culture of efficient strain of N_2 fixing organisms is grown on respective agar medium. A loopful of inoculums from it is transferred in a 300 ml of liquid medium. The flask is kept on shaker (260 rpm) for 72–96 hours. If shaker is not available, incubate it at 28°C for 5–6 days.

   (b) Preparation of broth culture:

   Each flask containing suitable broth is inoculated with the starter culture 1:2 proportion, aseptically. Incubate the flasks at 28°C for 2–5 days (Fig. 16.2), depending upon the type of organism till the count per ml reaches to 10^9 cells. This broth culture with population of 10^9 cell/ml should be used for preparation of carrier based biofertilizers.

Figure 16.2 Incubation of bioinoculant broth culture on rotary shaker.

16.3 Carrier-based biofertilizers

16.3.1 Preparation of carrier

Finely powdered peat, lignite or soil + compost or cellulose powder or soil + wood charcoals may be used as carrier. The carriers should have the following characteristics:

(1) High organic matter, above 60%

(2) Low soluble salt content (less than 1%)

(3) High moisture holding capacity (150–200% by weight)

Carrier provides a nutritive medium for growth of the bacteria and prolongs their survival in culture as well as on inoculated seed. The carriers are powdered to 250–300 mesh (about 75 micron pore size). Peat of 300 meshes is neutralized with 1% calcium carbonate and sterilized at 15 pounds p.s.i. for 4 hrs in autoclave.

16.3.2 Preparation of inoculants in powder form

In the preparation of inoculants, it is essential that the appropriate moisture content for a specific carrier should be determined. The ideal moisture range for the inoculant must be set while considering the bacterial growth and maximum population attainable after mixing, the bacterial survival and anticipated moisture loss from the package over the period of the shelf life.

Usually, about one part (by weight) of broth is required for two parts of dry carrier (Fig. 16.3). Final moisture content varies from 30–50%, depending on quality of carriers. After adding the broth culture to carrier powder in 1:2 proportion by weight, it is kept for curing at room temperature (28°C) for 5–10 days in 10 cm deep trays of convenient size. After curing, it is sieved to disperse the concentrated packets of growth and to break lumps. It is then packed in polythene bags of 0.5 mm thickness, leaving 2/3 space open for aeration of the bacteria (Figs. 16.4 and 16.5).

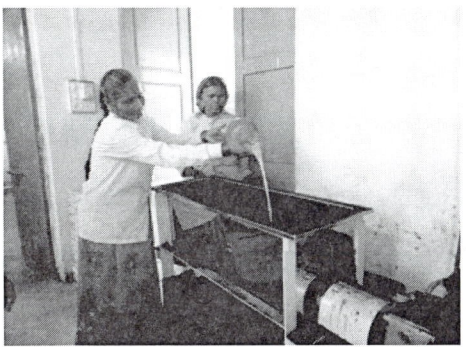

Figure 16.3 Mixing of broth culture in lignite carrier.

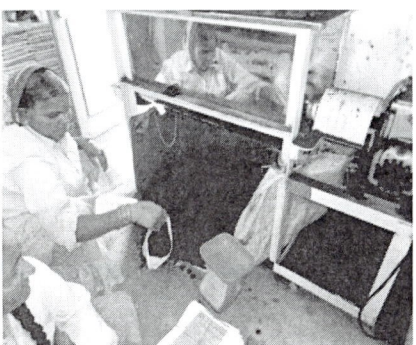

Figure 16.4 Packing of carrier-based biofertilizers.

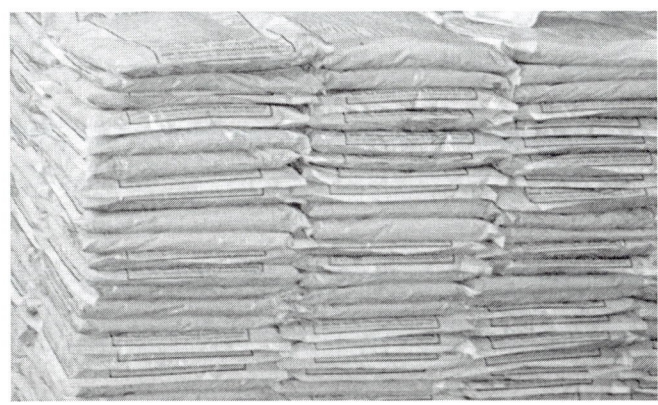

Figure 16.5 Carrier-based biofertilizer packets.

16.4   Liquid biofertilizer

16.4.1 Production technology of liquid biofertilizers

Materials:

Nutrient liquid media required by specific N_2 fixing organisms, i.e. Jensen's liquid medium for *azotobacter*, yeast extract mannitol broth for *rhizobium* and nitrogen-free broth for *azospirilium*, efficient cultures of diazotrophs, cell protectant, shaker, autoclaveble bottles, flasks, inoculating needle, bottling plant, etc.

16.4.2 Composition of liquid biofertilizer media

For *Rhizobium*

Mannitol	2.0 g	K_2HPO_4	0.5 g
$MgSO_47H_2O$	0.2 g	NaCl	0.1 g
Yeast Extract	0.5 g	Glycerol	4.0 ml
Trehalose	2.0 mM	Arbinose	5.0 g
PVP	20 g	Glucose	10.0 g
Fe-EDTA	200 ml	Distilled water	1000 ml

For *Azotobacter*

Sucrose	20.0 g	K_2HPO_4	0.5 g
$MgSO_47H_2O$	0.2 g	NaCl	0.1 g
Trehalose	2.0 mM	Glycerol	8.0 ml
PVP	1.5 g	Arbinose	0.5 g
Glucose	5 .0 g	Fe-EDTA	1.0 g
Yeast Extract	1.0 g	Trace element	2.0 ml
Malic acid	5.0 g	Solution	
Potassium Hydroxide	4.0 g	Ferric chloride	0.01 g
Calcium choloride	0.2 g	Polyvinyl acetate	3 %
Distilled water	1000 ml		

For *Azospirillum*

K_2HPO_4	0.5 g	Malic acid	5.0 g
$MgSO_47H_2O$	0.2 g	Sodium chloride	0.1 g
Sodiume molybdate	0.002 g	$MnSO_4$	0.2 g

KH₂PO₄	0.4 g	Ferric chloride	0.01 g
KOH	4.0 g	Bromothymol blue	5.0 ml
CaCl₂	0.2 g	(5% alchholic solution)	
Yeast Extract	1.0 g	Trehalose	2.0 mM
Glycerol	6.0 ml	PVP	1.5 g
Arbinose	5.0 g	Glucose	5.0 g
FeCl₃	0.01 g	Fe-EDTA	1.0 g
Distilled water	1000 ml		

For *Acetobacter*

Di potassium hydrogen orthophosphate	0.2 g
Potassium diydrogen phosphate	0.6 g
Magnesium sulphate	0.2 g
Sodium chloride	0.02 g
Sodium molybdate	0.002 g
Ferric chloride	0.01 g
Bromothymol blue	5 ml
Sucrose (Crystralized cane sugar)	100 g
Trehalose	2.0 mM
PVP (Polyvinyl phyrolidone)	1.5 g
Glycerol	8 ml
Distilled water	1000 ml

For Potash solubilizer

Magnesium sulphate	0.2 g
Sodium chloride	5.0 g
Trehalose	2.0 mM
Glycerol	8.0 ml
PVP	1.5 g
Glucose	10 g
Arabinose	0.5 g
Ferric chloride	0.1 g
Calcium carbonate	2.0 g
Distilled water	1000 ml

For PSB

Glucose	10.0 g	NaCl	5.0g
$MgSO_4 7H_2O$	0.2 g	Beef extract	1.5 g
Trehalose	2.0 mM	Glycerol	8.0 ml
PVP	1.5 g	Arbinose	0.5 g
$FeCl_3$	0.1 g	$CaCl_2$	0.1 g
NH_4Cl	0.5 g	Peptone	2.5 g
Distilled water	1000 ml		

16.4.3 Procedure for liquid biofertilizer production

(a) *Preparation of starter culture*: Pure culture of efficient strain of N_2 fixing organisms/potash solubilizer/phosphorus solubilizer is grown on respective agar medium. A loopful of inoculums from it is transferred in a 250 ml of liquid medium. The flask is kept on shaker (260 rpm) for 72–96 hours. If shaker is not available incubate it at 28°C for 5–6 days.

(b) Preparation of liquid biofertilizer:

Broth for respective efficient strain of N_2 fixing organisms/potash solubilizer/ phosphorus solubilizer is to be prepared. Cell protectant viz. trehalose, PVP is dissolve separately and added into the broth before sterilization. The sterilization of the broth is to be carried out at 15 lb/sq inch for 15 min. Saturated stock solution of glucose and arabinose is to be prepare, sterilized separately and added into the sterilized broth prepared earlier. This broth is inoculated with pure starter culture at 10 ml/lit under aseptic conditions. Incubate the flasks at 28°C for 2–5 days, depending upon the type of organism till the count per ml reaches 10^9 cells. Sterilized glycerol is to be added to the inoculated broth. This broth is to be dispensed in to the previously sterilized polypropylene bottles and made it airtight by screw cap. The bottles should provide the proper product identification specifying the strains for which it may be used, detailed application instructions for the user, an expiry date, quantity of seed or area to be treated and the manufacturing company.

Sulfur oxidizer and silicate solubilizer biofertilizer can be prepared by following the same procedure. The broth media prescribed for the respective microbial culture has to be used in each case.

Now-a-days, commercial liquid biofertilizer unit with automatic bottle filling machines are available (Fig. 16.6).

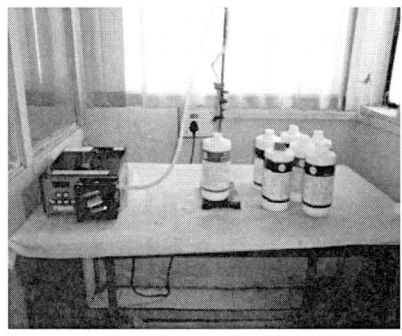

Figure 16.6 Commercial liquid biofertilizer unit with automatic bottle filling.

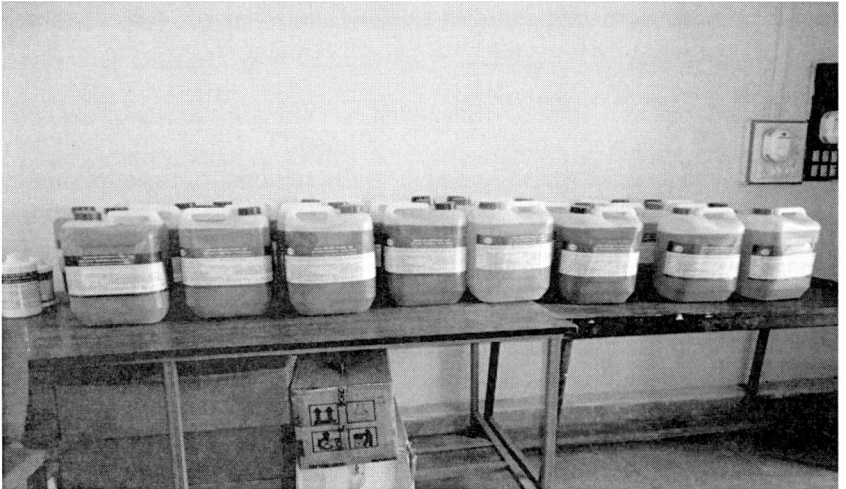

Figure 16.7 Liquid biofertilizer can.

(c) Advantages of liquid biofertilizers

- They have longer shelf life from 6 to 12 months.

- They are cost effective, pollution free and renewable source to supplement chemical fertilizers.

- They do not have effect of high temperature.

- They maintain high *cfu* more than 10^9/ml up to 12 months.

- They better survive on seeds and soil.

- Easy to use, handling and storage by farmers.

- The dosage of liquid biofertilizers is 10 times less that of carrier based biofertilizers.

- They have high export potential.

- Liquid biofertilizers can be packed in different volumes (Fig. 16.6).

16.5   Production technology for blue-green algae

The major factor limiting the use of Algal biofertllizer technology in large areas is the production of quality inoculum. A tested method is described below which is recommended for setting up large-scale production farms.

16.5.1 Mass production of BGA

Out of the different methods recommended for the production of BGA, open-air or controlled condition cement tank method (Fig. 16.7) is considered to be easy and cost effective. The cement tanks are permanent structures and can be cleaned easily. On the other hand, the galvanized iron sheet trays are very expensive, prone to rusting and difficult to clean. Similarly, the polythene-lined pits do not last for more than 3–4 harvests and thus become costly.

Production of BGA in cement tank:

(1) Construct cement tanks of size 5 m length, 1.5 m width and 0.3 m depth in an open space. The inner walls and floor of the tank should be glazed smooth. Provide a water tap at about 25 cm height at one of the broad sides and a drainage pipe fitted with a stop cock on the opposite wall at the bottom. An over flow outlet is provided at about 20 cm height above the drainage outlet. The lengths of the tanks can be manipulated, but the width should not be increased. Provide a space of about 1.0 m between the tanks for operational convenience.

(2) The four corners of the tank should be round.

(3) Spread 10 kg soil and add 200 g of single superphosphate per tank. The soil should be loam or sandy loam and not heavy type like black cotton or clay. Heavy soil can be mixed with appropriate quantity of riverbed sand. Light soil facilitates formation of algal flakes which can be easily separated out from the soil. The soil should be taken from a clean and fallow land, so that it has low nitrogen and microbial content.

(4) Fill the tanks with water up to a height of about 15 cm and add insecticide (10–15 ml malathion, 50% EC or endosulphan, 30% EC) to prevent breeding mosquito and other insects. Mix the contents thoroughly and allow to stand till the supernatant becomes clear.

(5) Sprinkle 200 g of good quality soil-based starter culture on the water surface.

(6) Under favorable conditions (temperature 30°C and above), the growth of blue-green algae will be rapid and thick algal mat will be formed on the surface of the water in about 10–15 days. At this stage, formation of another BGA layer can be seen on the surface of the soil. During this period, add water periodically to maintain the water level around 10 cm.

(7) Closely monitor the BGA coming up in the tank by periodically examining the algal growth using a microscope. One can alternatively use the iodine test to differentiate between blue-green algae and green algae. The green algae turn dark violet or black in color with iodine.

(8) Alkaline condition with pH around 8 appreciably prevents contamination.

(9) Stop adding water to the ponds only after a thick BGA mat is formed and allow the contents to dry without draining the water.

(10) When completely dry, the algal mat will form flakes which will separate out from rest of the soil. These flakes are collected; sun-dried and packed in polythene bag.

(11) Fill the tank again with water, put fresh soil, starter culture and superphosphate and repeat the process. Single harvest from a tank yields about 7–10 kg of soil-based algal flakes.

(12) Instead of using the soil-based algal culture, one can use laboratory-grown liquid culture to inoculate the tanks. Similar procedure is used for production of BGA in iron trays and soil beds (Fig. 16.8) using fresh soil, starter culture, and superphosphate.

Figure 16.8 Production of BGA in cement tank under controlled condition.

Figure 16.9 Production of BGA in open space in soil beds and iron trays.

16.5.2 Recommendation for field application

• Broadcast the dried algal flakes on standing water at the rate of 10–20 kg per hacter, one week after transplantation. Addition of excess quantity of algal flakes accelerates the multiplication and establishment of algae in the field. The field should be kept waterlogged for about 10 days after inoculation to allow good growth of BGA.

• If nitrogen fertilizers are not being used, apply BGA in order to gain a benefit of 15–25 kg N/ha. Where the nitrogenous fertilizers are used, keep the basal dose unchanged and reduce the subsequent two applications by half. This will save on the chemical fertilizers without affecting the yield.

• Apply blue-green algae at least for 3–4 consecutive seasons to have the advantage of cumulative effect.

• The sun-dried BGA inoculum packed in polythene bags can be stored for at least 3 years without loss in the viability.

• Recommended pest control measures and other crop management practices do not interfere with the establishment and activity of the BGA in the field.

• Application of small quantity of phosphatic fertilizer after BGA inoculation accelerates the algal establishment.

Precautions:

(1) The inoculum packets should be kept away from pesticides and fertilizers in a dry place.

(2) In Northern plains, BGA multiplication can be practiced during March–October.

(3) Do not allow insect breeding in the tanks.

(4) The algal material should be completely dried in the sun before packing.

(5) Construct the tanks in such a way so that the length of the pond is at right angle to the normal wind direction to have maximum light to allow more growth of algae.

(6) Use only single super phosphate to supply phosphorus. Do not add any nitrogen-containing fertilizer in the tank, since it will accelerate the growth of non-nitrogen fixing BGA and green algae.

(7) The quantity of water to be added to the tank should be standardized, depending upon the local conditions, temperature and rate of evaporation. Use of excess water may delay the harvesting of BGA flakes.

(8) Ensure the formation of thick algal mat on the water surface and another layer on the soil surface before allowing the formation of flakes.

(9) Do not drain the water to hasten the process of drying.

Advantages:

(1) Heavily fertilized rice fields show initial dominance of green algae even after algalization. These are replaced by the inoculated BGA as fertilizer nitrogen is consumed.

(2) The effect of BGA biofertilizer is not as instantaneous as in the case of inorganic fertilizers. It is slow, but sustainable and additive.

(3) Establishment of the algal inoculum in the field can be seen in the form of floating algal biomass on the surface of water or numerous small and glistening air bubbles adhering to the soil surface. These indications are best seen in the afternoon.

(4) After about 15–25 days of inoculation, the rice plants in the algalized plots appear greener than in the non-algalized plots.

(5) Algalization includes early grain setting and maturity which is indicated by the early drooping down of the ear-heads of the treated plants.

16.5.3 Importance of Algal biofertilizers

Nitrogen is a limiting nutrient in crop production. Despite its abundance in the air, higher plants are unable to assimilate this molecular nitrogen. The blue-green algae (BGA) is capable of metabolizing the elemental nitrogen through the process of biological nitrogen fixation. In tropical rice fields, biological nitrogen fixation is predominantly algae mediated process. Inoculating rice fields with selected isolates of blue-green algae adds 15–25 kg N/ha/season. The BGA is important input in rice field due to following reasons:

(1) Rice field conditions favor growth and nitrogen fixation by the blue-green algae.

(2) The BGA maintain a continuous supply of crop nutrients.

(3) BGA application improves the soil health.

(4) Continuous application for 3–4 consecutive seasons results in an appreciable population build up of the desired BGA in the soil.

(5) The inoculum can be easily produced by the farmer.

(6) Sun-dried BGA inoculum can be stored for over 3 years at ambient conditions.

(7) It is economical and easily adaptable by the small and marginal farmers.

16.6   Production technology of *Azolla*

For large-scale multiplication of *Azolla* in the field, the farmers have evolved a simple nursery method. The field selected for *Azolla* nursery should be thoroughly prepared and leveled uniformly. The field is divided into one-cent plot (20 × 20 m) by providing suitable bunds and irrigation channels. Water is maintained at a depth of 10 cm. Ten kilos of fresh cattle dung mixed in 20 liters of water is sprinkled in each plot and 8 kg of *Azolla microphylla* inoculum is introduced to each plot. Fifteen days after inoculation, *Azolla* is harvested from the main field as a source of primary inoculum. From one harvest, 80–100 kg fresh *Azolla* is obtained from each plot. Again, *Azolla* is inoculated in the same plot and cattle dung, superphosphate and furdan are applied. Four cents nursery (four bonded plots of 20 × 20 m) in flooded rice soil system is reported to increase the growth of *Azolla* and nitrogenase activity. The use of cow dung and cattle shed water increase the growth of *Azolla* during winter months in South India. The *Azolla* multiplied in four cent nursery is sufficient for 1 hecter of main rice field.

16.7   On-farm production of arbuscular mycorrhizal fungi

The on-farm production of arbuscular mycorrhizal fungi given by Verma (2011) is to be used. The arbuscular mycorrhizal fungal spores isolated from root rhizosphere of fruit crop (mango) are taken in 50 ml sterile water.

The required number of earthen pots are disinfected by 5% aqueous solution of copper sulphate and filled with fine sterilized soil.

The seeds of five trap crops viz. maize, berseem, fenugreek, marigold and sorghum are treated with sodium hypochlonide (2%) solution for 2 minutes and sown at appropriate depth. 50 ml sterile water containing arbuscular mycorrhizal fungal spores is added at the point of seed sown and covered with a layer of fine sterile soil over it. These pots are watered, as and when required.

The production of VAM can also be done in a field. The field of 25 m² is selected. Slurry is made from VAM fungal spore + sterile soil by adding sufficient quantity of water. This slurry is used for infection of AM fungi in trap crops. The seeds of trap crops are mixed with slurry of AM fungi spores and are sown in the field.

At the end of each crop cycle, the shoots of trap plants are removed from the soil surface without disturbing the roots; the roots are to be refined in the bed, but the soil should be hoed or loosened during second cycle of multiplication and so on. After five multiplication cycles, the plants are uprooted gently to get maximum root biomass. The roots with adhering soil are then chopped finely and thoroughly mixed. This serves as a source of inoculums of AM fungi.

16.8   Production technology of decomposting culture

16.8.1 Materials

Jowar grains, mother cultures of cellulolytic microorganisms and plastic bags.

16.8.2 Procedure

(1) Soak jowar grains in water for 24 hours.

(2) Sterilize the presoaked jowar grains for 30 minutes.

(3) Inoculate the grains with mother cultures of cellulolytic microorganisms at 10%.

(4) Allow the cellulolytic cultures to grow for 7 days. Use plastic tanks or a closed room for this purpose, when a large scale commercial production is to be done.

(5) Turning of the grains is done after 7 days to have good growth of microbe throughout the grain medium.

(6) Dry the jowar grain grown cellulolytic culture in shed for 4–5 days. Pack 500 g of culture in plastic bags

The rate and application of decomposing culture is one kg/ton of organic waste

16.9   Production of compost

16.9.1 Materials

Culture of cellulolytic microorganisms, urea, agro-residues, cow dung, single super phosphate and water, etc.

Pit method and heap method of compost preparation is followed (Fig. 29 and 30). However, pit method is widely followed over heap method.

16.9.2 Procedure

(1) Dig a pit approximately 3 feet deep and 6–8 feet wide near cattle shed on a site free from water logging.

(2) Cut plant residues into shreds of about 15–20 cm in length.

(3) Bring down C:N ratio of plant residue by sprinkling aqueous solution of urea (1 kg urea + 50 lit water per ton organic matter) on plant residues. For sugarcane trash, use aqueous solution of urea in high concentration (8.75 kg urea ton^{-1} of trash + 50 lit of water).

(4) Fill the shredded plant residue in a pit in layers (each layer approximately of about 30 cm).

(5) Apply mixture of cellulolytic fungi, *Trichoderma viride, Trichurs spiralis, Paecelomyces Fusisporus* and *Aspergillus* (10^6 cells g^{-1} of carrier) at 0.5 kg t^{-1} of organic matter, along with slurry (prepared in 50 lit of water by using dung at 10 % of organic matter).

(6) Sprinkle water to maintain optimum moisture.

(7) Repeat the process of layering and spreading cellulolytic culture till the residues reach 1 ft above the ground.

(8) Cover the compost pit with mud.

(9) Turn the composting material monthly to maintain proper aeration. Also sprinkle water at each turning to maintain moisture. Good quality compost will be ready within 3–4 months, depending upon the nature of the matter used.

16.9.3 Precautions

(1) As far as possible, prepare small pieces of residues.

(2) Spread cow dung slurry, aqueous solution of urea and decomposting culture evenly over each layer.

(3) Sprinkle suspension of urea first, followed by cow dung slurry and culture.

(4) Do not mix culture in urea.

(5) Maintain 60% moisture in the pit.

Figure 16.10 Pit method of compost preparation.

Figure 16.11 Heap method of compost preparation.

Assessment of quality standards of biofertilizers

17.1 Introduction

To achieve all the benefits from the use of biofertilizers, it is necessary to have a quality biofertilizer. The reasons for the poor quality of biofertilizers are use of ineffective microbial strain, insufficient number of viable cells, presence of contaminants, production by unskilled staff, poor shelf life and inadequate storage facilities. These factors if taken care of, the quality biofertilizers would be available to the farming community. It will help to improve the yield potential of crop plants and also in improving soil fertility. A number of brands of "biofertilizers" are in market and they have been found to vary in quality. Hence, there is an urgent need for Indian standards, not only to test the quality of inoculants, but also to provide the farmers with certified inoculants and to help the producers to improve the quality of their products.

17.2 Quality standards of *Rhizobium* and Azotobacter

Indian Standard Institution (Now Bureau of Indian Standards) formulated an agriculturally useful microorganisms sectional committee, which in 1977 specified the Indian standard for *Rhizobium* inoculants (IS: 8268–1976) and *Azotobacter* inoculants (IS: 9138–1979). These specifications are shown in Table 17.1.

Table 17.1 ISI standards specified for *Rhizobium* and *Azotobacter* biofertilizers

Parameters	*Rhizobium* Biofertilizer	*Azotobacter* Biofertilizer
Cell no. at the time of manufacture	10^8/g carrier within 15 days of manufacture	10^7/g carrier within 15 days of manufacture
Cell no. at the time of expiry date	10^7/g carrier within 15 days before expiry date	10^6/g carrier within 15 days before expiry date

Expiry date..	6 months from the date of manufacture	6 months from the date of manufacture
Permissible contamination level	No contamination at 10^8 dilurion	No contamination at 10^7 dilurion
pH	6.0–7.5	6.5–7.5
Strain	Should be checked serologically	Nothing specific. But *A. chroococcum* species is mentioned.
Carrier	Should pass through 150–212 microns IS sieve	Should pass through 160 microns IS sieve
Nodulation test	Should be positive	—
Nitrogen fixation	Above 20 mg/g of glucose	Not less than 10 mg/g of sucrose

The quality standards specified for *Rhizobium* in various countries are as follows (Table 17.2).

Table 17.2 Quality standards of commercial *Rhizobium* culture in different countries

Country	Cells/gm of culture (total viable count on Congo-red agar)		
	Very satisfactory	**Satisfactory**	**Doubtful**
U. S. A.	10^9	–	$10^6 – 10^7$
Australia	–	2×10^8	$10^6 – 10^7$
U. S. S. R.	10^9	–	–
India	More than 10^9	$10^7–10^9$	Less than 10^7

Although quality control standards for biofertilizer *Azospirillum* and PSM has not been in force, the proposed standard specification of PSM and *Azospirillum* are given in Table 17.3 (http://rbdcjabalpur.dacnet.nic.in).

Table 17.3 Proposed standard specifications of PSM and *Azospirillum*

S. No.	Parameter	PSM	*Azospirillum*
1	Base	Carrier (Lignite/ Charcoal)	Carrier (Lignite/ Charcoal)
2	Carrier	>100 micron	>100 micron
3	pH	6.5–7.5	7.0–8.0
4	Moisture	35–40%	35–40%
5	Viable count at manufacture	10^7/g carrier	10^7/g carrier

6	Viable count at expiry	10^7/g carrier	10^7/g carrier
7	Level of contaminant	Nil at 10^4 dilution	Nil at 10^4 dilution
8	Growth in Pikovskaya medium	+ve	–
9	Growth in S. S. Malate medium	–	+ve
10	P. Solubilization zone	1mm	–
11	Pellicle formation	–	+ve
12	Shelf life	6 months	6 months
13	P. Solubilization	30–50%	–
14	N-fixation	–	15 mg/g of malic acid

However, no regulatory act has been formulated/proposed in India to observe the quality control of biofertilizers like blue-green algae, *Azolla*, compost cultures and potash, sulphur and silicate solubilizing biofertilizers.

17.3 Quality control measures (as per ISI specification)

The biofertilizer should be assessed for the following quality standards:

(1) Inoculant should be carrier based or liquid based.

(2) The inoculant should contain minimum of 10^8 viable cells of bioinoculant/g of carrier on dry weight basis when it is stored at 25–30°C.

(3) The inoculant should have a maximum of 6 months of expiry period from manufacture in case of carrier based and 9 months in case of liquid based.

(4) The pH of inoculant should be between 6.0 and 7.5.

(5) Inoculant should show effective nodulation/nitrogen fixed on particular crop before expiry date.

(6) The carrier material should be in the form of powder, i.e. peat, lignite, peat soil, and humus, etc.

(7) Inoculant should be packed in 50–75 microns low-density polythene bags.

(8) Each packet should be marked legibly to give the information about name of the product, name of microbial inoculants, activity of bioinoculant, crop on which intended, name and address of manufacturer, type of carriers, batch and code numbers, date of manufacture, date of expiry, net quantity meant for 0.4 hectare, and storage instructions.

(9) It should be free from any contaminant/contamination with other microorganisms.

Application technology of biofertilizers

18.1 Use of carrier-based biofertilzers

There are basically three methods that are generally followed for the application of biofertilizers.

- Seed treatment
- Seedling treatment
- Soil treatment

18.1.1 Seed treatment

The technique for the application of carrier-based biofertilizers for seed treatment involves the following steps:

(1) Prepare the slurry of 250 g of biofertilizer in 200–500 ml water.

(2) Pour this slurry slowly on 10–15 kg of seeds, or seeds required for one acre land. Mix the seeds with hands evenly to get uniform coating of biofertilizers on all of them.

(3) Dry the treated seeds in shade and then sow immediately.

18.1.2 Seedling treatment

For seedling-transplanted crops, the seedling treatment of biofertilizers is as follows:

(1) Prepare the suspension of 1–2 kg of biofertilizer in 10–15 liters of water.

(2) Dip the roots of seedlings obtained from 250–500 seeds into the suspension for 20–30 minutes.

(3) Transplant the seedlings immediately.

18.1.3 Soil treatment

When biofertilizer application to seeds or seedling is not possible, the soil method of application is followed. The steps are as follows:

(1) Prepare the mixture of 2–4 kg of biofertilizer in 40–60 kg of soil compost.

(2) Broadcast the mixture in one acre of land either at sowing time or 24 hours beforesowing.

(3) For fruit crops, 5 kg FYM + 25 gm *Azotobacter* + 25 gm PSB + 25 gm Trichoderma is applied per plant.

Precautions

Follow the following precautions while using biofertilizer:

(1) Keep biofertilizer packets in a cool place away from direct heat/ sunlight.

(2) Mix biofertilizer with seeds as per requirement.

(3) Dry the treated seeds in shade.

(4) Sow treated seeds immediately.

(5) Do not use packets on which details regarding crop, date, batch number, name of the manufacturer and expiry date are not mentioned.

(6) Do not mix biofertilizer with insecticides, fungicides, herbicides, or chemical fertilizers.

(7) Use the *Rhizobium* inoculants specified for the particular crop.

18.2   Use of liquid biofertilizer

Methods of application of liquid bioinoculants

18.2.1 Seed Treatment

For 1 kg seed material, 25 ml of liquid bioinoculant is used. The seeds are dipped in this liquid for 10 minutes. Dry the seeds in shade and then sow them as early as possible, preferably during morning and evening hours. For seeds with hard seed coat like cotton, dipping should be carried out overnight.

18.2.2 Sett inoculation

The setts or tubers are immersed in liquid biofertilizers for 20 minutes and are planted in the field.

18.2.3 Seedling treatment

Root system of the seedlings is to be dipped in bioinoculant so that root system gets a high population of bioinoculant. The liquid bioinoculant (500 ml) is sufficient for seedling treatment for one acre area. Roots of seedlings are dipped in this biofertilizer solution for 8–10 min and transplanted immediately.

18.2.4 Soil treatment

If the seed treatment with liquid bioinoculant is not possible due to some circumstances, then 2 liter liquid bioinoculant is mixed with 50 kg of farmyard manure or compost. This mixture is spread uniformly on the field before irrigation. It can also be supplied to the soil through drip irrigation system.

18.2.5 Soil broadcasting

If the seed treatment with liquid bioinoculant is not possible, then the liquid bioinoculant can be applied to the crop plants by this method. First dilute 100 ml of liquid bioinoculant in 5 lit of water and then mixing with 50 kg of cow dung and 5 kg of rock phosphate. Keep this mixture overnight, and next day apply the mixture over one acre of land at root zone and irrigate.

18.2.6 Seed pelleting

Take 2 kg of sieved soil. Sprinkle 25 ml of liquid bioinoculant solution on sieved soil. Keep this mixture overnight. Take about 7–10 kg of seeds and mix with the mixture. Take care that seed coat remains intact. Allow the seeds to dry in shade before sowing in field.

18.2.7 Foliar spray

Dilute 3 liter of liquid bioinoculants in 200 liter water and spray this solution on sugarcane plant preferably in the evening.

18.28 Drip irrigation

Two liters of liquid biofetilizer is given through drip irrigation for one acre area.

Advantages, limitations and constraints

19.1 Advantages of biofertilizers

(1) Biofertilizers increase soil fertility by improving physical properties. Plant growth promoting substances and vitamins liberated by biofertilizers help to maintain soil health.

(2) They are cost effective, pollution free and a renewable source to supplement chemical fertilizers.

(3) Enrichment of soil with effective strains is beneficial for increasing crop yields.

(4) Biofertilizers are supplement to chemical fertilizers as they contribute plant nutrients through biological nitrogen fixation and solubilization of fixed phosphate, potassium, sulphur, or silica.

(5) They control and suppress soil borne diseases as some of the inoculants produce antibiotics.

(6) They help in root proliferation and survival of beneficial microbes in soil.

(7) They have a permanent effect.

(8) They reclaim the soils of high pH (about 12.5) by reducing it to pH 7.5. Thus, biofertilizers (blue-green algae) acts as a buffer for soil reaction.

(9) Biofertilizer (*Azotobacter*) helps to increase drought tolerance in plants.

(10) Biofertilizer (blue-green algae) has a wider range of pesticidal tolerance in plants.

(11) Can be used as green manure (e.g., *Azolla*) because of its large biomass and high nitrogen content.

(12) Helps to break the seed dormancy e.g., *Azotobacter*.

(13) Weed suppression is possible e.g., *Azolla*.

(14) Increased NO_3-reductase activity and antifungal compounds e.g., *Azotobacter.*

19.2 Limitations

(1) Indian Standard Institute (ISI) specifications are available only for *Rhizobium* and *Azotobacter.* Neither regulatory act nor facilities for testing the samples exist at present.

(2) No extension and propaganda of biofertilizers is being undertaken on large scale.

(3) Consumption of biofertilizers has not increased due to low level of farmer acceptance, probably because sub-standard materials are sold in the market.

(4) Their shelf life is only 6 months, and it is difficult under Indian conditions to transport, store and distribute the material in time. When the packets arrive in villages, they are either spoiled or over dated; and therefore, they become useless because organisms contained in biofertilizers die very rapidly in hot weather (40°C).

(5) As the soil dries, the population of bacteria (*Azotobacter*, Azospirillum, Rhizobium) decline drastically.

(6) Nitrogen-fixing organisms are adversely affected by low soil pH (<5.0), which is often associated with aluminum and manganese toxicity and calcium deficiency.

(7) Molybdenum is required for nitrogenase enzyme system during nitrogen fixation by *Azotobacter.*

(8) Numerous predators, parasitoids, and soil microorganisms hinder the establishment of biofertilizers.

19.3 Constraints in biofertilizer production technology

At present, there are several constraints in biofertilizer production and its commercialization. They may be physical, chemical, biological, technological, infrastructure, financial, market related, or concerned with human resource development.

19.3.1 Physical and environmental constraints

In summer months in some parts of India temperature is very high. Care must be taken to avoid transportation of biofertilizer packets in direct sunlight. Storage of biofertilizers should not be done in containers/sheds where the temperature becomes very high. While making packets, care must be taken to see that the carrier material has enough moisture (about 40%). If the carrier material is very dry, population of organisms decreases rapidly.   ⁻⁻

19.3.2 Chemical constraints

Highly acidic or saline soils will adversely affect the population of introduced biofertilizer. In acidic soil, pelleting of seeds (after treating with biofertilizers) with lime and in saline soils with gypsum can be recommended. In soils where P availability is less than 10 ppm, application of P-fertilizers is essential without which the nitrogen-fixing biofertilizers will not function effectively.

19.3.3 Biological constraints

Soil flora and fauna can parasitize or devour biofertilizer added to soil. Bacteriophages can destroy rhizobia added to soil. Some protozoa like *verticelia* and nematodes like *Meloidogyne* can act as predators on biofertilizers added to soil. *Azolla* is subjected to attack of pests, especially larvae of various lepidopterous and dipterous insects and aphids. Cynobacteria can be attacked by phages, myxobacteria, and grazers like snails and mosquito larvae.

19.3.4 Technical constraints

Raw material

Peat and lignite which are good source of carrier materials are available only in very limited places like Nilgiris and Neyveli, thus resulting in high cost of transportation to biofertilizer manufacturing units.

Strains

Suitable strains of biofertilizers for different crops have been identified through research. Many producers do not pay attention to this aspect and use whatever strains they get, resulting in poor quality biofertilizers.

Suitable technology

In India, biofertilizer production is generally done with non-sterile carrier which encourages considerable contamination. Suitable technology to minimize contamination should be developed, e.g., gamma irradiation of carrier material, injection of liquid culture into polythene bags containing sterile carrier material, etc. Recently, there has been some interest to produce liquid biofertilizers. They are special liquid formulations containing not only the desired microorganisms and their nutrients but also special cell-protectants or chemicals that promote formation of resting spores or cysts for longer shelf life and tolerance to adverse conditions (Bhattacharya and Kumar 2002).

The liquid biofertilizer of good quality holds great promise for the following benefits over carrier material:

(1) Saving on carrier material

(2) Transport

(3) Pulverization and sterilization

(4) Convenience in handling

(5) Storage and transportation and better performance.

(6) Wider applicability

(7) More expiry period

Lack of quality assurance

Substandard quality of inoculants is one of the most important factors resulting in failure at the field and lack of farmer's confidence in product. The bureau of Indian standards prepared ISI specification for *Rhizobium* and *Azotobacter* inoculants in 1977. Revised specification for *Rhizobium* and *Azotobacter* and specification for *Azospirillum* and phosphobacteria have been formulated recently.

Quality control consists of testing of bioinoculant strains for their effectiveness and number of microbial cells in broth before incorporation into carrier materials like lignite, charcoal, peat, vermiculite, etc., so as to get the appropriatiable cell number at the time of manufacturing and control of culture during the period of storage; therefore, each individual manufacturer has to do his own quality control. The BIS has no network to test biofertilizers samples. They send samples to agricultural universities or research institutes in order to make ISI mark. Once manufacturer gets an ISI mark, they continue using it without routine checking. This renders the substandard product from reaching to the farmers. It is suggested that there should be separate testing laboratories with adequate infrastructure and manpower to check the quality of biofertilizers at various stages of production, marketing and application.

19.3.5 Infrastructure constraints

Many production units do not have adequate space and other supplies needed for quality biofertilizers.

19.3.6 Human constraints

Some biofertilizer production units do not have technically well-trained persons (microbiologists) who can produce high quality biofertilizers. It has been experienced that some organizations depend more on non-skilled labor on contract basis, which leads to substandard biofertilizer production. Most of the farmers in our country do not have sufficient and clear knowledge about the use of biofertilizers. The farmers need to be educated. In order to educate farmers, providing education and training to workers is important. This can be done through demonstration trials on the cultivator's fields. The impact is greater and long lasting when farmers see the beneficial results for themselves on their lands.

An intense publicity programme is required for motivating the people creating an assurance and disseminating knowledge amongst farming community. This can be done through media like T.V., cassettes, radio and advertisements in newspaper, seminars, exhibitions and rural fairs.

ADP (Adenosine diphosphate): A compound which upon phosphorylation (addition of phosphate and energy) forms high energy bonds as ATP.

Aerobe: An organism which grows in presence of oxygen.

Agar: A gelatin-like material obtained from seaweed, and used to prepare culture media on which microorganisms are grown and studied.

Anaerobe: An organism which grows in the absence of oxygen.

Antibiotic: A chemical compound produced by one microorganism which inhibits or kills other microorganisms.

Arbuscule: A branched, ruft-like haustorium, produced by certain mycorrhizal fungi inside root cells.

Associative symbiosis: The loose association between the roots of non-leguminous crop and N-fixing bacteria.

ATP (Adenosine triphosphate): A compound formed by phosphorylation of ADP and which stores and releases energy for the various cell function.

Autotroph: Organism able to utilize carbon dioxide as sole source of carbon.

Bacteria: The prokaryotic organisms.

Bacteriods: Irregular and enlarged forms of rod shaped bacteria, especially of the root nodule forming species.

Capsule: A relatively thick layer of mucopolysaccharides which surrounds some kinds of bacteria.

Cell: The fundamental unit of life.

Cellulose: A polysaccharide composed of hundreds of glucose molecules linked in a chain and found in the plant cell walls.

Cellulolytic microorganisms: The microorganisms which have the ability to degrade plant residues (cellulose-rich materials) in soil.

Colony: A contagious group of single cells derived from a single ancestor and growing on a solid surface.

Culture: Growth of microorganisms on artificial media.

Curing: Process of establishment of microbes in artificial carrier.

Cyst: An encysted spore (bacteria).

Diazotrophs: The organisms which fix atmospheric nitrogen.

DNA (Deoxyribonucleic acid): The genetic material of organisms.

Encyst: To form a cyst.

Endotrophic mycorrhizae: Fungus that forms an association with woody and herbaceous plants, penetrating the cortex of the roots both inter and intracellularly.

Eubacteria: The spherical, ovoid or rod shaped unicellular microorganisms which multiply by transverse, binary fission.

Enzyme: A protein produced by living cells that can catalyze a specific organic reaction.

Fermentation: Oxidation of certain organic substances in the absence of molecular oxygen.

Fission: Transverse splitting in two of bacterial cells.

Flagellum: A whip like structure projecting from a bacterium and functioning as an organ of locomotion.

Habitat: The natural place of occurrence of an organism.

Heterocyst: Specialized cells on the algal filament in which nitrogen fixation takes place. These are large, thick walled, apparently empty cells growing in between pigmented cells on the algal filament.

Heterotrophic: Depending on an outside source for organic nutrients.

Host: A plant that is invaded by a parasite and from which the parasite obtains its nutrients.

Hydrolysis: The enzymatic breakdown of a compound through the addition of water.

Incubation: Holding cultures of microorganisms under condition favorable for their growth.

Inoculation: The transfer of mother culture into the broth medium.

Inoculum: The organism or its parts that can cause infection.

Isolate: A single spore or culture and the subcultures derived from it.

Isolation: The separation of an organism from its host and its culture on a nutrient medium.

Micro-aerophylic: Organisms which are able to grow only at low dissolved oxygen.

Microorganisms: Microscopic organisms within the categories algae, bacteria, fungi (including lichen), protozoa, viruses and sub viral agents; also see organism.

Micron (μ): A unit of length equal to 1/1000 of a millimeter.

Millimicron (mμ): A unit of length equal to 1/1000 of a micron.

Millimeter (mm): A unit of length equal to 1/10 of a centimeter (cm).

Metabolism: The process by which cells or organisms utilize nutritive material to build living matter and structural components, or break down cellular material into simple substances to perform special functions.

Mycorrhiza: A symbiotic association of a fungus with the roots of a plant.

Nitrification: The oxidation of ammonium to nitrate in soil by bacteria.

Oxidation: A chemical reaction in which oxygen combines with another substance or in which hydrogen atoms or electrons are removed from a substance.

pH: It is a measurement of the degree of acidity or alkalinity of a solution.

Photosynthesis: The process by which carbon dioxide and water are combined in the presence of light and chlorophyll to form carbohydrate.

Polysaccharide: A large organic molecule consisting of many units of a simple sugar.

Prokaryote: A microorganism whose genetic material is not organized into a membrane-bound nucleus.

Rhizosphere: The soil near a living root.

RNA (Ribonucleic acid): A nucleic acid involved in protein synthesis.

Shelf life: The period up to which biofertilizer contains a certain minimum specified number of viable bacteria

Slime layer: Diffused and extensive layer present (secreted) outside the bacterial cell; not a part of the bacterial cell.

Sterilization: The elimination of pathogens and other living organisms from soil, containers, etc., by means of heat or chemicals.

Starter culture: Pure culture used for starting a fermentation process.

Strain: The descendants of a single isolation in pure culture; an isolate. Also, a group of similar isolates.

Sub-culture: Transfer of culture or portion of culture to fresh nutrient medium.

Symbiosis: A mutually beneficial association of two different kinds of organisms.

Unicellular: Consisting of one cell.

Viability: Term applied to population to denote the ratio of viable organisms to the total number of organisms present.

Viable: Capability of organism to germinate, live, and grow.

(1)   Ashby's medium

Mannitol	10.0 g	K_2HPO_4	0.2 g
$MgSO_47H_2O$	0.2 g	NaCl	0.2 g
K_2SO_4	0.2 g	K_2SO_4	0.1 g
Distilled water	1000 ml	pH	7.0

(2)   Jensen's medium

Sucrose	20.0 g	K_2HPO_4	1.0 g
$MgSO_47H_2O$	0.5 g	NaCl	0.5 g
$FeSO_4$	0.1g	$CaCO_3$	2.0 g
Agar	15.0 g	Distilled water	1000 ml
pH	7.0		

(3)   Yeast extract mannitol agar medium

Mannitol	10 g	K_2HPO_4	0.5 g
$MgSO_47H_2O$	0.2 g	NaCI	0.1 g
Yeast extract	1.0 g	Distilled water	1000 ml
Agar	20.0 g	Congo red (1%)	2.5 ml
pH	7.0		

(4)   Sodium malate medium/NFB medium

K_2HPO_4	0.5 g	Malic acid	5.0 g
$MgSO_47H_2O$	0.1 g	Sodium chloride	0.02 g
Sodium molybdate	0.002 g	$MnSO_4$	0.01 g

KH$_2$PO$_4$	0.4 g	FeSO$_4$ 7H$_2$O	0.05 g
KOH	4.0 g	Bromothymol blue (5% alcoholic solution)	2.0 ml
CaCl$_2$	0.01 g	Distilled water	1000 ml
Agar	1.75 g	pH	6.8

(5) Pikovskaya's medium

Glucose	10.0 g	Ca$_2$ (PO$_4$)$_2$	5.0 g
MgSO$_4$7H$_2$O	0.1 g	KCl	0.2 g
(NH$_4$)$_2$ SO$_4$	0.5 g	MnSO$_4$	trace
FeSO$_4$7H$_2$O	trace	Yeast extract	0.5 g
Agar	15.0 g	Distilled water	1000 ml

(6) LGIP medium

Sucrose	100.0 g	K$_2$HPO$_4$	0.4 g
MgSO$_4$7H$_2$O	0.2 g	KH$_2$PO$_4$	0.6 g
FeCl$_2$	0.01g	CaCl$_2$	0.02 g
Sodium molybdate	0.02 g	Agar	2.0 g
Distilled water	1000 ml	Bromothymol blue (5% alcoholic solution)	5.0 ml
pH	5.5		

(7) Diluted cane juice semisolid medium

Semisolid LGIP medium	250 ml
Sugarcane cane juice	250 ml
Distilled water	500 ml

(8) Semisolid acetified LGIP medium

Semisolid acetified LGIP medium was acidified with acetic acid to pH 4.5 and agar concentration was increased to 2.2 g l^{-1}.

(9) Hoyer's medium

(NH$_4$)$_2$SO$_4$	1.0 g

KH$_2$PO$_4$	0.9 g
MgSO$_4$7H$_2$O	0.25 g
K$_2$HPO$_4$	0.1g
Fecl$_3$6H$_2$O	0.02g
Ethanol solution	200 ml

Add ethanol to distilled water and bring volume to 200 ml. Mix thoroughly and filter sterilize.

Add all the components, except ethanol solution, in water and bring volume to 800 ml, autoclave and cool to add ethanol solution.

(10) Nutrient agar medium

Peptone	5.0 g	Beef extract	3.0 g
Sucrose	20.0 g	Agar	20.0 g
Distilled water	1000 ml	pH	7.0

(11) Sulphur enriched medium

(NH$_4$)$_2$SO$_4$	0.2 g	KH$_2$PO$_4$	3.0 g
MgSO$_4$7H$_2$O	0.5 g	FeSO$_4$7H$_2$O trace	
CaCl$_2$	0.2 g	Elemental sulphur	10.0 g
Distilled water	1000 ml		

(12) Starkey's medium for thiobacillus thiooxidants

Elemental sulphur	10.0 g
Potassium dihydrogen phosphate	4.0 g
Ammonium sulphate	0.4 g
Magnesium sulphate	0.5 g
Calcium chloride	0.25 g
Ferric sulphate	0.01 g
Distilled water	1000 ml
pH	2.0–3.5

After sterization of medium, about 0.25 gm of sterile sulphate is dusted on to the surface of the medium.

(13)   Thiosulphate agar for thiobacillus thioparus

Sodium thiosulphate	5.0 g
Sodium biocarbonate	0.2 g
Dipotassium phosphate	0.1 g
Ammonium chloride	0.1 g
Agar	15.0 g
Distilled water	1000 ml

(14)   Thiosulphate medium

Sodium thiosulphate	5.0 g	Sodium nitrate	3.0 g
K_2HPO_4	1.0 g	$MgSO_47H_2O$	0.5 g
KCl	5.0 g	$FeSO_47H_2O$	0.01 g
Sucrose	20.0 g	Agar	15.0 g
Distilled water	1000 ml		

(15)   Cellulolytic media for fungi (asparagines medium)

$(NH_4)_2SO_4$	0.5 g	L-Asparagine	0.5 g
$MgSO_47H_2O$	0.2 g	$CaCl_2$	0.1 g
KH_2PO_4	1.0 g	KCl	0.5 g
Cellulose	10.0 g	Yeast extract	0.5 g
Distilled water	1000 ml	pH	6.2

(16)   Cellulolytic media for bacteria (Han's medium)

$(NH_4)_2SO_4$	1.0 g	$MgSO_47H_2O$	0.1 g
$CaCl_2$	0.1 g	KH_2PO_4	0.5 g
NaCl	6.0 g	K_2HPO_4	0.5 g
Cellulose	10.0 g	Yeast extract	0.1 g
Distilled water	1000 ml	pH	6.5

(17)   Frankia broth

Yeast extract	5.0 g	Dextrose	10.0 g
Casamino acid	5.0 g	Vitamin B	0.001 g
Distilled water	1000 ml	pH	6.4

(18) Fogg's medium

KH_2PO_4	0.2 g	$MgSO_47H_2O$	0.2 g
$CaCl_2$	0.1 g	Na_2MOO_4	0.1 mg
$MgCl_2$	0.1 mg	H_3BO_3 (boric acid)	0.1 mg
$CuSO_4$	1.0 mg	$ZnSO_4$	0.1 mg
FE-EDTA	2.0 ml	Distilled water	1000 ml

References

- Abd-Elmonem, E. A. and Amberger, A. (2000). Studies on Some Factors Affecting the Solubilization of P from Rock Phosphate, 6th international collegum for the optimization of plant nutrition, Cairo, Egypt.

- Ahmad, F., Ahmad, I., and Khan, M. S. (2005). Indole Acetic Acid Production by the Indigenous Isolates of Azotobacter and Fluorescent Pseudomonas in the Presence and Absence of Tryptophan, *Turkish journal of biology* (**29**), pp. 29–34.

- Ahmed, M. and Ahmed, N. (2007). Genetics of Bacterial Alginate: Alginate Genes Distribution, Organization and Biosynthesis in Bacteria, *Current Genomics*, **8**(3), pp. 191–202.

- Allaby, M. (Editor) (1992). Algae, The concise Dictionary of Botany, Oxford Universtiy Press.

- Allen, E. B. (1989). The Restoration of Disturbed Arid Landscapes with Special Reference to Mycorrhizal Fungi, *Journal of arid environment*, **17**, pp. 279–286.

- Aleksandrov, V. G., Blagodyr, R. N., and Liiev, I. P. (1967). Liberation of Phosphoric Acid from Apatite by Silicate Bacteria, *Mikrobiyol* Zh. (Kiev), **29**, pp. 111–114.

- Al-Sherif, E. M. (1998). Ecological Studies on the Flora of Some Aquatic Systems in Beni-Suef District, M. Sc. Thesis, Beni-Suef, Cairo University (Beni-Suef branch), Egypt.

- Aragno, M. (1991). Aerobic Chemolithoautotrophic Bacteria, In: Thermophilic bacteria, Kristjansson, J. K. (Editor) (CRC Press, BocaRaton, Fla), pp. 7–103.

- Armstrong, D. C. (1988). Role of phosphorus in plants in: Better Crops with Plant Food, PP4-5 ed. DI Armstrong (Atlanta USA: potash and phosphate institute).

- Arnold, C. A. (1955). A Tertiary *Azolla* from British Columbia, contributions from the Museum of Paleontology, University of Michigan, **12**(4), pp. 37–45.

- Arveby, A. and Huss-Danell, K. (1988). Presence and Dispersal of Infective Frankia in Peat and Meadow Soils in Sweden, Biology and Fertility of Soils, **6**, p. 1007.

- Avakyan, Z. A., Pivovarova,T. A., and Karavaiko, G. I. (1986). Properties of a New Species, Bacillus mucilaginosus, Mikrobiol, **55**, pp. 477–482.

- Badr, M. A. (2006). Efficiency of K-feldspar Combined with Organic Materials and Silicate Dissolving Bacteria on Tomato Yield, *Journal of Applied Sciences Res.*, **2**(12), pp. 1191–1198.

- Bagyaraj, D. J. (1989). Response of Crop Plants to VAM Inoculation in an Unsterile Indian Soil, New photologist, **35**, pp. 33–36.

- Baillie, A., Hodgkiss, W., and Norris, J. R. (1962). Flagellation of *Azotobacter* spp. as Demonstrated Electron Microscopy, *Journal of Applied Microbiology*, **25**(1), pp. 116–119.

- Baker, D. and David, O'Keefe (1984). A Modified Sucrose Fractionation Procedure for the Isolation of Frankia from Actinorhizal Root Nodules and Soil Samples, Plant and soil, **78**, pp. 23–28.

- Barbosa, Heloiza, R., Marcos A. Moretti, Daniela, Thuler, S., and Elisabeth, Augusto, F. P. (2002). Nitrogenase activity of Beijerinckia Derxii is preserved under adverse conditions for its growth. *Brazilian Journal of Microbiology*, **33** (3).

- Bashan, Y. and Levanony, H. (1990). Current Status of *Azospirillum* Inoculation Technology: *Azospirillum* as a Challenge for Agriculture, *Canadian Journal of Microbiology*, **36**, pp. 591–608.

- Bashan, Y. (1993). Potential Use of *Azospirillum* as Biofertilizer, Turrialba, **43**, pp. 286–291.

- Bashan, Y. and Holguin, G. (1994). Applied and Environmental Microbiology, Root-to-root ravel of the beneficial bacterium *Azospirillum brasilense*, **60**, pp. 2120–2131.

- Bashan, Y. and Holguin, G. (1995). Microbial Ecology, Inter-root movement of *Azospirillum brasilense* and subsequent root colonization of crop and weed seedlings growing in *soil*, **29**, pp. 269–281.

- Bashan, Y., Puente, M. E., Rodriguez-Mendoza, M. N., Toledo, G., Holguin G., Ferrera-Cerrato, R., and Pedrin, S. (1995). Applied and

Environmental Microbiology, Survival of *Azospirillum brasilense* in the bulk soil and rhizosphere of 23 soil types, **61**, pp. 1938–1945.

- Becking, J. H. (1959). Nitrogen Fixing Bacteria of the Genus Beijerinckia in South African Soil, *Plant and soil*, **11**(3), pp. 193–206.

- Becking, J. H. (1961). Nitrogen Fixing Bacteria of the Genus Beijerinckia, *Soil Science*, 118, pp. 196–212.

- Beijerinck, M.W. (1901). Ueber Oligonitrophile Mikroben, *Zentralblatt fur Bakteriologie Parasitenkunde, Infektionskrankheiten Han Hygiene*, Abteilung II (In German), **7**, pp. 561–582.

- Bellenger, J. P., Wicharz, T., and Kraepiel, A. M. L. (2008). Vanadium Requirments and Uptake Kinetics in the Dinitrogen-Fixing Bacterium Azotobacter vinelandii, *Applied and Environmental Microbiology*, **74**(5), pp. 1478–1484.

- Ben Dekhil, S., Cahill, M., Stackebrandt, E., and Sly, L. I. (1997). Transfer of *conglomeromonas largomobilis sub. sp. largomobilis* to the genus Azospirillum as *Azospirillium largomobile comb*. Nov and elevation of *conglomeromonas largomobilis subsp parooensis* to the new type species of conglomeromonas, *conglomeromonas parooensis sp. Nov. Syst Applied Microbiology*, **20**, pp. 72–77.

- Berg, R. H. (1990). Cellulose and Xylans in the Interface Capsule in Symbiotic Cells of Actinorhizae, *Protoplasma*, **159**, pp. 35–43.

- Berkum, P. V., and Bahlool, B. B. (1980). Evaluation of Nitrogen Fixation by Bacteria in Association with Roots of Tropical Grasses, *Microbiology Rev.*, **44**, pp. 491–517.

- Berthelin, J. (1983). Microbial Weathering Processes, In Microbial Geochemistry. Krumbein, W. E. (Editor), Blackwell Scientific Publications, pp. 223–262.

- Bertsch, P. M., and Thomas, G. W. (1985). Potassium Status of Temperate Region Soils, In: Munson, R. D. (Ed.) Potassium in agriculture ASA, CSSA, and SSSP, Madison, WI, pp. 131–162.

- Bethlenfalvay, G. J., Franson, R. L., Brown, M. S., and Mihara, K. L. (1989). The Flycine-Flomus-Bradyrhizobium symbiosis IX, Nutritional, morphological, and physiological responses of nodulated soybean to geographic isolates of the mycorrhizal fungus Glomus mosseae, *Physiol. Plant*, **76**, pp. 226–232.

- Bocchi, S., and Malgioglio, A. (2010). Azolla-Anabaena as a Biofertilizer for Rice Paddy Fields in the Po Valley, a Temperate Rice Area in Northern Italy, *International Journal of Agronomy*.

- Boddey, M. R., and Dobereiner, J. (1994). Association of *Azospirillum* and other Diazotrophs in the Tropical Gramineae, *Transactions of the 12th International Congress of Soil Science*, New Delhi Symposia Papers I, pp. 28–47.

- Bowen, C. D., Skinner, M. F., and Bevege, D. I. (1974). Zinc Uptake by Mycorrhizal and Uninfected Roots of Pinus Radiate and Arauaria Cumninghami, *Soil biology and biochemistry*, **6**(3), pp. 141–144.

- Boyd, W. L., and Boyd, J. W. (1962). Presence of *Azotobacter* Species in Polar Regions, *Journal of Bacteriology*, **83**(2), pp. 429–430.

- Brinkhuis, H., Schouten, Schouten, S., Collinson, M. E., Sluijs, A., Sinninghe Damste, J. S., Dickens, G. R., Huber, M., Cronin, T. M., Onodera, J., and Takahashi K. (2006). Episodic Fresh Surface Waters in the Eocene Arctic Ocean, *Nature*, **441**(7093), pp. 606–609.

- Brown, M. E. (1972). Plant Growth Substances Produced by Microorganisms of Soil and Rhizosphere, *Journal of Applied Bacteriology*, **35**, pp. 443–451.

- Brune, D. C. (1989). Sulfur Oxidation by Phototrophic Bacteria, *Biochim. Biophys. Acta*, **975**, pp. 189–221.

- Brune, D. C. (1995). Sulfur Compounds as Photosynthetic Electron Donors, In: Anoxygenic Photosynthetic Bacteria, Blankenship, R. E., Madigan, M. T., Bauer C. E. (Editors) (Kluwer, Dordrecht, The Netherlands), pp. 847–870.

- Bumb, B. (1995). Global Fertilizer Perspective, 1980–2000: The Challenges in Structural Transformation, *Technical bulletin T-42*, Muscle schoals, AL: International fertilizer development centre.

- Burgmann, H., Widmer, F., Sigler, W. V., and Zeyer, J. (2003). mRNA Extraction and Reverse Transcription-PCR Protocol for Detection of nifH Gene Expression by *Azotobacter vinelandii* in Soil, *Applied and Environmental Microbiology*, **69**(4), pp. 1928–1935.

- Burris, R. H. (1994). Biological Nitrogen Fixation-past and Future, In: Hegazi, N. A., Fayez, M., Monib, M. (Editors), Nitrogen fixation with non-legumes, Cairo, Egypt, The American University in Cairo Press, pp. 1–11.

- Callaham, D., Newcomb, W., Torrey, J. G., and Peterson, R. L. (1979). Root Hair Infection in Actinomycete Induced Root Nodule Initiation in Casuarinas, Myric and comptonia, Bot Gaz (Chicago), **140** (Supp), pp. 51–59.

- Carrapico, F., and Tavares, R. (1989a). New Data on the *Azolla*-Anabaena Symbiosis, I-Morphological and Histochemical Aspects, In

Nitrogen Fixation with non-legumes, Skinner, F A. et. al. (Editors), Kluwer Academic Publishers, pp. 89–94.

- Carrapico, F., and Tavares, R. (1989b). New Data on the *Azolla*-Anabaena Symbiosis, II-Cytochemical and Immunocytochemical Aspects, In Nitrogen Fixation with Non-legumes, Skinner, F. A. et. al. (Editors), Kluwer Academic Publishers, pp. 95–100.

- Carrapico, F. (1991). Are Bacteria the Third Partner of the *Azolla*-Anabaena symbiosis? *Plant and Soil,* **137**, pp. 157–160.

- Carrapico, F., Teixeira, G., and Diniz, M. A. (2000). Azolla as Biofertilizer in Africa, A Challenge for the Future, *Revista de Ciencias Agrarias*, **23**(3–4), pp. 120–138.

- Cavalcante, V. A., and Dobereiner, J. (1988). A New Acid Tolerant Nitrogen Fixing Bacterium Isolated with Sugarcane, *Pl. Soil*, **108**, pp. 23–31.

- Chandra, K., Greep, S., Ravindranath, P., and Sivathsa, R. S. H. (2005). Liquid Biofertilizers, Regional Center for Organic Farming Hebbal, Bangalore.

- Chein, S. H., Menon, R. G., and Billingham, K. (1996). Phosphorus Availability from Phosphate Rock as Enhanced by Water-soluble Phosphorus, *Soil- sci. Soc. American Journal*, **60**, pp. 1173–1177.

- Chen, J. H., Czajka, D. R., Lion L. W., Shuler, M. L., and Ghiorse, W. C. (1995). Trace Metal Mobilization in Soil by Bacterial Polymers, *Environmental Health Perspectives*, **103**(1), pp. 53–58.

- Chinnasami, K. N., and Chandrasekaran, S. (1978). Silica Status in Certain Soils of Tamil Nadu, *Madras Agric. J.*, **65**, pp. 743–746.

- Christophe, C., Turpault, M. P., and Freyklett, P. (2006). Root Associated Bacteria Contribute to Mineral Weathering and to Mineral Nutrition in Trees and Budgeting Analysis, *Applied Environmental Microbiology*, **72**(2), pp. 1258–1266.

- Ciobanu, I. (1961). Investigation on the Efficiency on Bacterial Fertilizers Applied to Cotton, *Cent. Exp. Ingras. bact. Lucrari. Stiint.*, **3**, pp. 203–214.

- Clarson, D. (2004). Potash Biofertilizer for Eco-friendly Agriculture, Agro-clinic and Research Centre, Poovanthuruthu, Kottayam (Kerala), India, pp. 98–110.

- Cohen. Y, Jorgensen, B. B., Revsbech, N. P., and Poplawski, R. (1986). Adaptation to Hydrogen Ulfide of Oxygenic and Anoxygenic

Photosynthesis among Cyanobacteria, *Applied Environmental Microbiology*, **51**(2), pp. 398-407.

- Cohen, M. F., Meziane, T., Suchiya, M., and Yamasaki, H. (2002). Feeding Deterrence of *Azolla* in Relation to Deoxyanthocyanin and Fatty Acid Composition, *Aquatic Bot.*, **74**(2), pp. 181–187.

- Costa, M. L., Carrapico, F., and Santos, M. C. R. (1994). Biomass and Growth Characterization of *Azolla filiculoides* in Natural and Artificial Environments, In Nitrogen Fixation with Non-legumes, Hegazi, N. A., Fayez, M., and Monib, M. (Editors), The American University in Cairo Press, pp. 455–461.

- Cumpkin, T. A., and Plucknett, D. L. (1982). Azolla as a Green Manure: Use and Management in Crop Production, West view Press, Boulder, Colorado, p. 230.

- Curatti L., Brown, C. S., Ludden, P. W., and Rubio, L. M. (2005). Genes required for rapid expression of nitrogenase activity in Azotobacter vinelandii, *Proceedings of the National Academy of Sciences of the United States of America*, **102**(18), pp. 6291–6296.

- Datta, M., Banish, S., and Gupta, R. D. (1982). Studies on the Efficacy of a Phytohormone Producing Phosphate Solubilizing Bacillus Firmus in Augmenting Paddy Yield in Acid Soils of Nagaland, *Plant and Soil*, **69**, pp. 365–373.

- Datta, N. P., and Shinde, J. E. (1985). Yield and Nutrition of Rice Under Upland and Waterlogged Conditions, Effect of nitrogen, phosphorus and silica, *J. Indian Soc. Soil Sci.*, **33**, pp. 53–60.

- De los Rios, A., Grube, M., Sancho, L. G., and Ascaso, C. (2007). Ultrastructural and Genetic Characteristics of Endolithic Cyanobacterial Biofilms Colnizing Antartic Granite Rocks, *FEMS Microbiology Ecology*, **59**(2), pp. 386–95.

- De Zwart J. M. M., Nelisse, P. N., and Kuenen, J. G. (1996). Isolation and Characterization of *Methylophaga Sulfidovorans*, sp. Nov.: An obligately methylotrophic, aerobic, dimethyl sulfide oxidizing bacterium from a microbial mat, *FEMS Microbiol. Ecol.*, **20**, pp. 261–270.

- Dela Cruz, R. E., Manalo, M. Q., Aggangan, N. S., and Tambalo, J. D. (1988). Growth of Three Legume Trees Inoculated with VA Mycorrhizal Fungi and Rhizobium, *Plant and Soil*, **108**, pp. 111–115.

- Denison, R. F. (2000). Legume Sanctions and the Evolution of Symbiotic Cooperation by Rhizobia, *American Naturalist*, **156**, pp. 567–576.

- Derx (1950). Beijerinckia, A New Genus of Nitrogen Fixing Bacteria Occurring in Tropical Soils, Proc. Koninklijke Nederlandse-Akademie van wetenschappn series c, **53**, pp. 140–147.

- Dicker, H. J., and Smith, D. W. (1980). Enumeration and Relative Importance of Acetylene-Reducing (Nitrogen-Fixing) Bacteria in a Delaware Salt Marsh, *Applied and Environmental Microbiology*, **39**(5), pp. 1019–1025.

- Diem, H. G., and Dommergues, Y. R. (1984). Isolation, Characterization and Cultivation of Frankia, In: Biological Nitrogen Fixation Recent Developments, Edited by Subha Rao, N. S., Gordon and Breach Science Publication, **1988,** pp. 225–244.

- Dixon, R. O. D., and Wheeler, C. T. (1986). Nitrogen Fixation in Plants, Glasgow, United Kingdom, Blackie.

- Dobereiner, J. (1959). Nitrogen Fixation Associated with Non-legumes Plants, In: Genetic Engineering for Nitrogen Fixation, Hollaender, A. (Editor), Plenum Press, New York, pp. 451–461.

- Dobereiner, J. (1961). Nitrogen Fixing Bacteria in the Rhizosphere, In: The Biology of Nitrogen Fixation, Quispel, A. (Editor), North Holland, Amsterdam, pp. 86–120.

- Dobereiner, J. (1961). Nitrogen Fixing Bacteria of the Genus Beijerinckia in the Rhizosphere of Sugarcane, *Plant of Soil*, **15**, pp. 211–216.

- Dobereiner, J. (1966). Azotobacter paspali sp. n. una bacteria fixadora de nitrogenio na rizosfera de paspalum, Pesquisa Agropecuaria Brasileira, **1**, pp. 357–365.

- Dobereiner, J., Day, J. M., and Dart, P. J. (1972). Nitrogenase Activity and Oxygen Sensitivity of the *Penicillium notatum* and *Azotobacter paspali* association, *J. Gen. Microbiol.*, **71**, pp. 103–116.

- Dong, C. T., Liu, Z. G., Zou, B. J., Zhu, C., Zhang,, C. L., and Liang, W. (1981). Study on the Nutrition of Rice (2), Effect of Zinc and Silicon in Increasing Rice Yield, *Lianing Agric. Sci.*, **4**, pp. 13–18.

- Drozdowicz, A., and Ferreira Santos, G. M. (1987). Nitogenase Activity in Mixed Cultures of *Azospirillum* with other Bacteria, *Zentralblatt far Mikrobiologie*, **142**, pp. 487–493.

- Durrant, M. C., Francis, A., Lowe, D. J., Newton, W. E., and Fisher, K. (2006). Evidence for a Dynamic Role for Homocitrate during Nitrogen Fixation: The Effect of Substitution at the α-Lys[426] Position in MoFe

Protein of *Azotobacter* vinelandii, *Biochemistry Journal*, **397**(2), PP. 261–270.

- Eckert, B., Weber, O. B., Kirchhof, G., Halbritter, A., Stoffels, M., and Hartmann, a. (2001). *Azospirillum doebereinerae* sp. Nov., a nitrogen fixing bacterium associated with the C4-grass Miscanthus, *Int J. Syst Evol Microbiol*, **51**, PP. 17–26.

- Ehrlich, H. L. (1981). Geomicrobiology, Marcel Dekker Inc., Newyork, p. 393.

- Elawad, S. H., Gascho, G. J., and Street, J. J. (1982). Response of Sugarcane to Silicate Source and Rate, I, Growth and Yield, *Agron. J.*, **74**, pp. 481–484.

- Emitiazia, G., Ethemadifara, Z., and Habibib, M. H. (2004). Production of Extra-cellular Polymer in *Azotobacter* and Biosorption of Metal by Exopolymer, *African Journal of Biotechnology*, **3**(6), pp. 330–333.

- Epstein, E. (1999). Silicon, Annual Review of Plant Physiology and Plant Molecular Biology, **50**, pp. 641–664.

- Eriksson, E. (1963). The Yearly Circulation of Sulphur in Nature, *J. Geophys. Res.*, **68**(13), pp. 4001–4008.

- FAO (1998). Guide to Efficient Plant Nutrition Management, FAO/ AGL publication, FAO, Rome.

- FAO (1990). Fertilizer Yearbook, Vol. **39**, Rome, Italy: FAO.

- Forni, C., Caiola, G., and Gentili, S. (1989). Bacteria in the *Azolla*-Anabaena Symbiosis, In Skinner, F. A. et. al. (Editors), Nitrogen Fixation with Non-legumes, Kluwer Academic Publishers, pp. 83–88.

- Francis, R., and Read, D. J. (1984). Direct Transfer of Carbon Between Plants Connected by Vesicular Arbuscular Mycorrhizal Mycelium, *Nature*, **307**, pp. 53–56.

- Friedrich, C. G., and Mitrenga, G. (1981). Oxidation of Thiosulfate by *Paracoccus denitrificans* and other Hydrogen Bacteria, *FEMS Microbiol. Lett.*, **10**, pp. 209–212.

- Friedrich, S. N., Platonova, P., Karaviko, G. I., Stichel, E., and Glombitza, F. (1991). Chemical and Microbiological Solublization of Silicates, *Acta Biotech*, **11**, pp. 187–196.

- Friedrich, C. G. (1998). Physiology and Genetics of Sulfur-oxidizing Bacteria, Adv. Microb. Physiol., **39**, pp. 235–289.

- Friend, J. P. (1973). The General Sulfur Cycle, In: Chemistry of lower atmosphere, (Rasool, S.) (Editor), pp. 177–201, New York, Plenum press.

- Fuchs, T., Huber, H., Burggraf, S., and Stetter, K. O. (1996). 16S rDNA-based Phylogeny of the Archaeal Order *Sulfolobales* and Reclassification of *Desulfurolobus ambivalens* as *Acidianus ambivalens* comb. Nov. Syst. *Appl. Microbiol.*, **19**, pp. 56–60.

- Fuentes- Ramirez, L. E., Jimenez-salgado, T., Abarca-ocampo, R., and Caballeco-Mellado, J. (1993). Acetobacter diazotrophicus, an Indoleacetic Acid Producing Bacterium Isolated from Sugarcane Cultivars in Mexico, *Plant Soil*, **154**, pp. 145–150.

- Funa, N., Ozawa, H., Hirata, A., and Horinouchi, S. (2006). Phenolic Lipid Synthesis by Type III Polyketide Synthases is Essential for Cyst Formation in Azotobacter vinelandii, *Proceedings of the National Academy of Sciences of the United States of America*, **103**(16), 6356–6361.

- Galindo, E., Pena, C., Nunez, C., Segura, D., and Espin, G. (2007). Molecular and Bioengineering Strategies to Improve Alginate and Polydydroxyalkanoate Production by Azotobacter Vinelandii, *Microbial Cell Factories*, **6**(7), p.7.

- Gallo, J. R., Furlani, P. R., Bataglia, O. C., and Hiroce, R. (1974). Silicon Content in Grass and Forage Crops, *Cienciae culture*, **26**, pp. 282–293.

- Gama-Castro, S., Nunez, C., Segura, D., Moreno, S., Guzman, J., and Espin, G. (2001). Azotobacter Vinelandii Aldehyde Dehydrogenase Regulated by c54: Role in Alcohol Catabolism and Encystment, *Journal of Bacteriology*, **183**(21), pp. 6169–6174.

- Gandora, V., Gupta, R. D., and Bhardwaj, K. K. R. (1998). Abundance of *Azotobacter* in Great Soil Groups of North-West Himalayas, *Journal of the Indian Society of Soil Science*, **46**(3), pp. 379–383.

- George Garrity, M. (Editor) (2005). Part B: The Gammaproteobacteria, *Bergey's Mannual of Systematic Bacteriology*, The Proteobacteria (2nd Edition), New York: Springer, ISBN 0-387-95040-0.

- Gerdemann, J.W. (1955). Wound-healing of Hyphae in a Phycomycetous Mycorrhizal Fungus, *Mycologia*, **47**, pp. 916–118.

- Gerdemann, J.W., and Nicolson, T. H. (1963). Spores of Mycorrhizal Species Extracted from Soil by Sieving and Decanting, *Transaction of the British Mycological Society*, **46**, pp. 235–244.

- Gibson, A. H., Scowcraft, W. R., Child, V. J., and Pagan, J. D. (1976). Nitrogenase Activity in Cultured Rhizobium JP. Strain 32H1, *Archives of Microbiology*, **108**(1), 45–54.

- Gillis, M., and J. De. Ley (1980). Intra and Intergeneric Similarities of the Ribosomal RNA Cistrons of Acetobacter and Gluconobacter, *Int. J. Systemic Bacteriol.*, **30**, pp. 7–27.

- Gillis, M., Kersters, K., Hoste, B., Janssens, D., Krappenstedt, R. M., Stephan, M. P., Teixeira, K. R. S., Dobereiner, J., and J. De Ley (1989). Acetobacter Diazotrophicus sp. nov., A Nitrogen Fixing Acetic Acid Bacterium Associated with Sugarcane, *Int. J. Syst. Bacteriol*, **39**, pp. 361–364.

- Giraud, Eric, L. et. al., Vallenet, D., Barbe, V., Cytryn, E., Avarre, J. C., Jaubert, M., and Simon, D. (2007). Legumes Symbioses: Absence of Nod Genes in Photosynthetic Bradyrhizobia, *Science*, **316**(5829), pp. 1307–12.

- Goldstein, A. H. (1994). Involvment of the Quino Protein Glucose Dehydrogenase in the Solubilization of Exogeneous Mineral Phosphates by Gram Negative Bacteria, In: Phosphate in Micro-organisms, *Cellular and molecular biology*, Eds., pp. 197–203.

- Granat, L., Rodhe, H., and Hallberg, R. O. (1976). The Global Sulphur Cycle, *Ecological bulletin* (Stockholm), **22**, pp. 89–134.

- Grey, E. A. (1953). Cotamination of Azotobacter Chroococcum by Gram Negative Bacterial Rod, *Nature*, **171**, p. 1163.

- Gromov, B. V. (1957). The Microflora of Rocks and Primitive Soil in Some Northern Regions of the USSR, *Mikrobiologiya*, **26**, pp. 52–54.

- Grover, R. (2003). Rock Phosphate and Phosphate Solubilizing Microorganisms as a Source of Nutrients for Crops, A report of Department of Biotechnology and Environmental Sciences, Thapar Institute of Engineering and Technology, Patiala, pp. 51.

- Halsall, D. M., and Gibson, A. H. (1989). Nitrogenase Activity of a Range of Diazotrophic Bacteria on Straw, Straw Breakdown Products and Related Compounds, *Soil Biology and Biochemistry*, **21**, pp. 291–298.

- Han, H. S., and Lee, K. D. (2005). Phosphate and Potassium Solubilizing Bacteria Effect on Mineral Uptake, Soil Availability and Growth of Eggplant, *Res. J. Agric. Biol. Sci.*, **1**(2), pp. 176–180.

- Han, H. S., Supanjani, and Lee, K. D. (2006). Effect of Co-inocualtion with Phosphate and Potassium Solubilizing Bacteria on Mineral Uptake and Growth of Pepper and Cucumber, *Pl. Soil Envrion.*, **52**(3), pp. 130–136.

- Hans Gunter Schlegel, Zaborosch, C., and Kogut, M. (1993). General Microbiology, Cambridge University Press, p. 380.

- Harley, J. L., and Smith, S. E. (1983). Mycorrhizal symbiosis, Academic press, New York, Chapter 5, pp. 104–116.

- Harrison, A. P. Jr. (1984). The Acidophilic Thiobacilli and Other Acidophilic Bacteria that Share their Habitat, *Annu. Rev. Microbiol.*, **38**, pp. 265–292.

- Heath, K. D., and Tiffin, P. (2009). Stabilizing Mechanisms in Legume-rhizobium Mutualism, *Evolution*, **63**(3), pp. 652–662.

- Heckman, J. (2006). A History of Organic Farming: Transitions from Sir Albert Howard's War in the Soil to USDA National Organic Program, *Renew. Agric. Food Syst.*, **21**, pp. 143–150.

- Heinen, W. (1960). Silicon Metabolism in Microorganisms, *Arkiv Mikrobiol*, **37**, pp. 199–210.

- Hill, S., Austin, S., Eydmann, T., Jones, T., and Dixon, R. (1996). Azotobacter Vinelandii NIFL is a Flavoprotein that Modulates Transcriptional Activation of Nitrogen-fixation Genes via a Redox-sensitive Switch, *Proceedings of the National Academy of Sciences of the United States of America*, **93**(5), pp. 2143–2148.

- Hiraishi, A., and Umeda, Y. (1994). Intragenic Structure of the Genus Rhodobacter transfer of Rhodobacter sulfidophilus and Related Marine Species to the Genus Rhodovulum gen. nov., *Int. J. Syst. Bacteriol.*, **44**, pp. 15–23.

- Holguin, G., and Bashan, Y. (1993). Increasing the Nitrogen Fixing Activity of *Azospirillum* by Mixed Culturing with Staphylococcus sp. In: New Horizons in Nitrogen Fixation (R. Palacios, J. Mora and W. E. Newton, Eds), p.726, Kluwer Academic Publishers, London.

- Hopper, W. (1993). Indian Agriculture and Fertilizer: An Outsiders Observation "Key Note Address to the FAI Seminar on Emerging Scenario in Fertilizer and Agriculture: Global Dimension, The Fertilizer Association of India, New Delhi.

- Howard, J. B., and Rees, D.C. (2006). How Many Metals Does it Take to Fix N2? A Mechanistic Overview of Biological Nitrogen Fixation, *Proceedings of the National Academy of Sciences of the United States of America*, **103**(46), pp. 17088–17093.

- Hu, X. F., Chen, J., and Guo, J. F. (2006). Two Phosphate and Potassium Solubilizing Bacteria Isolated from Tiannu Mountain, Zhejiang, China, *World J. Micro. Biotechnology*, **22**, pp. 983–990.

- Hu, Y., Fay, A. W., Lee, C. C., and Ribbe, M. W. (2007). P-cluster Maturation on Nitrogenase MoFe Protein, *Proceedings of the National Academy of Sciences of the United states of America*, **104**(25), pp. 10424–10429.

- Huber, R., and Stetter, K. O. (1999). Aquificales In: Embryonic encyclopedia of life sciences, London (Macmillan, Houndmills, England), pp. 1–7.

- Hung, R. S., Smith, W. K., and Yost, R. S. (1985). Influence of Vesicular-arbuscular Mycorrhiza on Growth, Water Relations Leaf Orientation in Leucaena Leucocephala, New Phytol., **99**, pp. 229–243.

- Hung, L. L., and Sylvia, D. M. (1988). Production of Vesicular-arbuscualr Mycorrhizal Fungus Inoculums in Aeroponic Culture, *Appl. Environ. Microbiol.*, **54**, pp. 353–357.

- Hussner, A. (2006). NOBANIS- Invasive Alien Species Fact Sheet-Azolla Filiculoides Online Database of the North European and Baltic Network on Invasive Alien Species, Heinrich Heine Univ. Dusseldorf. www.nobanis.org/files/factsheets/Azolla filiculoides.pdf.

- IFA: (2014). International fertilizer industry Association. http://www.fertilizer.org/ifa/Homepage/About.IFA.

- Imhoff J. F., Suling, J., and Petri, R. (1998). Phylogenetic Relationship and Taxonomic Reclassification of Chromatium Species and Related Purple Sulfur Bacteria, *Int. J. Syst. Bacteriol.*, **48**, pp. 1029–1043.

- Isopi, R., Fabbri, P., Del, Gallow, M., and Puppi, G. (1995). Dual Inoculation of Sorghum Bicolor (L.) Moench asp. Bicolor with Vesicular Arbuscular Mycorrhizae and *Acetobacter diazotrophicus*, *Symbiosis*, **18**, pp. 43–55.

- Jarstfer, A. G., and Sylvia, D. M. (1992). Inoculum Production and Inoculation Stratégies for Vesicular Arbuscular Mycorrhizal Fungi, In: Soil Microbial Ecology, Meting, B. (Editor), Applications in Agriculture and Environmental Management, Marcel Dikker, New York, pp. 349–377.

- Jaga, P. K., and Patel, Y. (2012). An Overview of Fertilizers Consumption in India: Determinants and Outlooks 2020-A review, *International J. of scientific enginerring & Technology*, **1**(6), pp. 285–291.

- James, E. K., Reis., V. M., Olivares, F. L., Baldani, J. I., and Dobereiner, J. (1994). Infection of Sugarcane by the Nitrogen-fixing Bacterium Acetobacter diazotrophicus, *Journal of Experimental Botany*, **45**(6), pp. 757–766.

- Jaspar, H. L., Jalali, B. L., and Varma, S. J. (1987). Vesicular Arbuscular Mycorrhiza: Progress and Projection, In: Mycorrhiza Round Table, Jawaharlal Nehru University, New Delhi, pp. 13–15.

- Jensen, H. L. (1954). The Azotobacteriaceae, *Bacteriological Reviews*, **18**(4), pp. 195–214.

- Jones, D. H. (1920). Further Studies on the Growth Cycle of *Azotobacter*, *Journal of Bacteriology*, **5**(4), pp. 325–341.

- Johstone, D. B. (1967). Isolation of *Azotobacter* insignis from Fresh Water, *Ecology*, **48**(4), pp. 671–672.

- Kass, D. L., Drosdoff, M., and Alexander, M. (1971). Nitrogen Fixation by Azotobacter paspali in Association with Bahia grass (*Paspalum notatum*), *Soil Science Society America Journal*, **35**(35), pp. 286–289.

- Khammas, K. M., and Kaiser, P. (1992). Pectin Ecomposition and Associated Nitrogen Fixation by Mixed Cultures of *Azospirillum* and Bacillus species, *Canadian Journal of Microbiology*, **38**, pp. 794–797.

- Kaneshiro, T., Crowell, C. D., and Hanrahan Jr., R. F. (1978). Acetylene Reduction Activity in Free Living Cultures of Rhizobia, *International J. of systematic and evolutionary microbiology*, **28**(1), pp. 27–31.

- Kannaiyan, S. (1986). Studies on Azolla pinnata for Rice Crop, *Research Journal of Plant Environment*, **3**, pp. 1–16.

- Kannaiyan, S. (2000). Biofertilizer Technology and Quality Control, Publication Directorate, Tamil Nadu Agricultural University, Coimbatore, India, p. 256.

- Kellogg, W. W., Cadle, R. D., and Allen, E. E. (1972). The Sulphur Cycle, Science, **175**, pp. 587–596.

- Kelly, D. P. (1982). Biochemistry of the Chemolithotrophic Oxidation of Inorganic Sulfur, Phil. Trans. R. Soc. Lond. *B. Biol. Sci.*, **298**, pp. 499–528

- Kelly, D. P. (1989). Physiology and Biochemistry of Unicellular Sulfur Bacteria, In: Autotrophic bacteria, Schlegel, H. G., Bowien, B. (Editors) (Springer, Berlin, Germany), pp. 193–217.

- Kelly, D. P. and Smith, N. A. (1990). Organic Sulfur Compounds in the Environment, Adv. Microb. Ecol., **11**, pp. 345–385.

- Kelly, D. P., Shergill, J. K., Lu, W. P., and Wood, A. P. (1997). Oxidative Metabolism of Inorganic Sulfur Compounds by Bacteria, Antonie Leeuwenhoek, **71**, pp. 95–107.

- Kelly, D. P., McDonald, I. R., and Wood, A. P. (2000). Proposal for the Reclassification of Thiobacillus Novellus as Starkeya Novella gen. nov., comb. Nov., in the -subclass of the Proteobacteria, *Int. J. Syst. Evol. Microbiol.*, **50**, pp. 1797–1802.

- Kelly, D. P. and Wood, A. P. (2000). Reclassification of some Species of *Thiobacillus* to the Newly Designated Genera *Acidithiobacillus* gen. nov., *Halbacillus* gen. nov. and *Thermithiobacillus* gen. nov., *Int. J. Syst. Evol. Microbiol.*, **50**, pp. 511–516.

- Keyser, H. H., and Li, F. D. (1992). Potential for Increasing Biological Nitrogen Fixation in Soybean, *Plant and Soil*, **141**, pp. 119–135.

- Khammas, K. M., Ageron, E., Grimont, P. A. D., and Kaiser, P. (1989). *Azospirillum* Irakense sp. Nov., A Nitrogen Fixing Bacterium Association with Rice Roots and Rhizosphere Soil, *Res. Microbiology*, **140**, pp. 679–693.

- Khudsen, D., Peterson, G. A., and Prov, P.F. (1982). Lithium, Sodium, and Potassium, In: Page, A. L. (Editor), Methods of soil Analysis Part (2) Agronomy Monograph, **9**, 2nd Edition, ASA and SSSA Madison, WI.

- Kiers, E.T., Rousseau, R. A., West, S. A., and Denison, R. F. (2003). Host Sanctions and the Legume-rhizobium Mutualism, *Nature*, **425**, pp. 79–81.

- Koide, R. T. and Mooney, R. P. (1987). Regulation of the VAM Symbiosis, *Annual review of Plant Physiology and Plant Molecular Biology*, **43**, pp. 557–581.

- Kondratieva, E. N. (1989). Chemolithotrophy of Phototrophic Bacteria, In: Autotrophic bacteria, Schlegel, H. G., Bowien, B. (Editors) (Springer, Berlin, Germany), pp. 283–287.

- Kovalcic, D. A., St. John, T. V., and Dyer, M. I. (1984). Lack of Vesicular Arbuscular Mycorrhizal Inoculums in a Pondersa Pine Forest, *Ecology*, **65**(6), pp. 1755–1759.

- Krajick, K. (1998). Archeology: Green Farming by the Incas, *Science*, **281**(5375), p. 322.

- Krieg, N. R. and Holt, J. G. (1984). In: Bergey's Manual of Systematic Bacteriology.

- Kumar, R., Bhatia, R., Kukregja, K., Behl, R. K., Dudeja, S. S., and Narula, N. (2007). Establishment of *Azotobacter* on Plant Roots: Chemotactic Response, Development and Analysis of Root Exudates of Cotton (Gossypium hirsutum L.) and Wheat (Triticum aestivum L.), *Journal of Basic Microbiology*, **47**(5), pp. 436–439. ˙˙

- Lalonde, M., and Quispel, A. (1977). Ultra Structural and Immunological Demonstration of the Nodulation of the European *Alnus glutinosa* (L.) Gaertn. Host Plant by the North-American *Alnus crispa var. mollis* Fern. Root nodule endophyte, *Can. J. Microbiol.*, **23**, pp. 1529–1547.

- Layne, J. S. and Johnson, E. J. (1964). Natural Factors Involved in the Induction of Cyst Formation in *Azotobacter*, *Journal of Bacteriology*, **87**(3), pp. 684–689.

- Lea-Smith, D. J., Ross, N., Zori, M., Bendall, D. S., Dennis, J. S., Scott, S. A., Smith, A. G., and Howe, J. (2013). Thylakoid Terminal Oxidases are Essential for the Cyanobacterium Synechocystis sp. PCC 6803 to Survive Rapidly Changing Light Intensities, *Plant Physiology*, **162**(10), PP. 484–495.

- Lee, M., Breckenridge, C., and Knowles, R. (1970). Effect of Some Culture Conditions on the Production of Indole-3-acetic Acid and Gibberellin like substances by Azotobacter vinelandii, *Can J. Microbial.*, **16**(12), pp. 1325–1330.

- Lee, R. E. (2008). Phycology, Fourth edition, Cambridge University Press, p. 534.

- Lewis, I. M. (1937). Cell Inclusions and the Life Cycle of Azotobacter, *Journal of Bacteriology*, **34**(2), pp. 191–205.

- Lewis, I. M. (1941). The Cytology of Bacteria, *Bacteriological Reviews*, **5**(3), pp. 181–230.

- Li, D. Y., Eberspacher, J., Wagner, B., Kuntzer, J., and Lingens, F. (1991). Degradation of 2,4,6-trichlorophenol by *Azotobacter* sp. Strain GP1, *Applied and Environmental Microbiology*, **57**(7), pp. 1920–1928.

- Li, Y. F. (1994). The Characteristics and Function of Silicate Dissolving Bacteria Fertilizer, *Soil and Fertilizer*, **2**, pp. 48–49.

- Libbert, E. and Risch, H. (1969). Interaction Between Plants and Epiphytic Bacteria Regarding their Auxin Metabolism, V, Isolation and Identification of IAA-producing and Destroying Bacteria from Pea Plants, *Physiol. Plan.*, **22**, pp. 51–58.

- Lin, L. P. and Sadoff, H. L. (1969). Preparation and Ultrastructure of the Outer Coats of Azotobacter vinelandii Cysts, *Journal of Bacteriology*, **98**(3), pp. 1335–1341.

- Lin, L. P., Pankratz, S., and Sadoff, H. L. (1978). Ultrastructural and Physiological Changes Occurring upon Germination and Outgrowth of Azotobacter vinelandii cysts., *Journal of Bacteriology*, **135**(2), pp. 641–646.

- Lin, Q. M., Rao, Z. H., Sun, Y. X., Yao, J., and Xing, L. J. (2002). Identification and Practical Application of Silicate Dissolving Bacteria, *Agric. Sci. China*, **1**, pp. 81–85.

- Loperfido, B. and Sadoff, H. L. (1973). Germination of Azotobacter vinelandii Cyssts: Sequence of Macromolecular Synthesis and Nitrogen Fixation, *Journal of Bacteriology*, **112**(2), pp. 841–846.

- Lowry, O. H., Rosebrough, N. J., Farr, A. F., and Randall, R. J. (1951). Protein Determination by Lowry's Method, *J. Biol. Chem.*, pp. 193–265.

- Macy, J. M., Schroder, I., Thauer, R. K., and Kroger, A. (1986). Growth of *Wolinella succinogenes* on H_2S plus Fumarate and on Formate Plus Sulfur as Energy Sources. *Arch. Microbiol.*, **144**, pp. 147–150.

- Magalhaes, F. M., Baldani, J. I., Souto, M., Kuykendall, J. R., and Dobereiner, J. (1983). A New Acid Tolerant *Azospirillum* Species, *Ann acad Bras Cienc*, **55**, pp. 417–430.

- Maier, R. J. and Moshiri, F. (2000). Role of the Azotobacter vinelandii Nitrogenase-Protective Shethna Protein in Preventing Oxygen-Mediated Cell Death, *Journal of Bacteriology*, **182**(13), pp. 3854–3857.

- Marco, D. E., Perez-Arnedo, R., Hidalgo-Perea, A., Olivares, J., El Ruiz-Sainz, J., and Sanjuan, J. (2009). A Mechanistic Molecular Test of the Plant-sanction Hypothesis in Legume-rhizobia Mutualism, *Acta Oecologica-International Journal of Ecology*, **35**, pp. 664–667.

- Martyniuk, S., and Maryniuk, M. (2003). Occurrence of Azotobacter Spp. in Some Polish Soil, *Polish Journal of Environmental Studies*, **12**(3), pp. 371–374.

- Michiels, K., Vanderleyden, J., and Van Gool, A. (1989). *Azospirillum*-plant Root Associations: A Review, *Biology and Fertility of Soils*, **8**, pp. 356–368.

- Mikhailouskaya, N. and Tcherhysh, A. (2005). K-mobilizing Bacteria and their Effect on Wheat Yield, *Latvian J. Agron.*, **8**, pp. 154–157.

- Miller, R. W., and Eady, R. R. (1988). Molybdeum and Vanadium Nitrogenases of Azotobacter Chroococcum, Low Temperature Favors N_2 Reduction by Vanadium Nitrogenase, *Biochemistry Journal*, **256**(2), pp. 429–432.

- Mishustin, E. N. and Shilinikova, V. K. (1972). Biological Fixation of Atmospheric Nitrogen by Free Living Bacteria, In: Soil Biology, Review of Research, UNESCO, Paris, pp. 82–109.

- Moira, K. D., Henderson, and Duff, R. B. (1963). The Release of Metallic and Silicates Ions from Minerals, Rocks and Soils by Fungal Activity, *J. Soil Sci.*, **14**, pp. 237–245.

- Money, T., Barrett, J., Dixon, R., and Austin, S. (2001). Protein-Protein Interactions in the Complex between the Enhancer Binding Protein NIFA and the Sensor NIFL from Azotobacter vinelandii, *Journal of Bacteriology*, **183**(4), pp. 1359–1368.

- Moore, A. W. (1969). Azolla: Biology and Agronomic Significance, *Botanical Review*, **35**, pp. 17–34.

- Moorman, T. B., and Reeves, F. B. (1979). The Role of Endomycorrhizae in Vegetation Practices in the Semi-arid West, II Bioassay to Determine the Effect of Land Disturbance on Endomycorrhizal Population, *Amer. J. Bot.*, **66**, pp. 14–18.

- Moreno J., Gonzalez-Lopez, J., and Vela, G. R. (1986). Survival of Azotobacter spp. in Dry Soils, *Applied and Environmental Microbiology*, **51**(1), pp. 123–125.

- Muralikannan, N. (1996). Biodissolution of Silicate, Phosphate, and Potassium by Silicate Solubilizing Bacteria in Rice Ecosystem, M. Sc. (Ag) Thesis Submitted to Tamil Nadu Agricultural University, Coimbatore, p. 125.

- Muralikannan, N. and Anthomiraj, S. A. (1998). Occurrence of Silicate Solublising Bacteria in Rice Ecosystem, *Madras Agric. J.*, **85**(1), pp. 47–50.

- Murry, M. A., Fontaine, M. S., and Torrey, J. G. (1984). Growth Kinetics and Nitrogenase Induction in Frankia sp. HFPArI 3 Grown in Batch Culture, *Developments in plant and soil science*, **12**, pp. 61–78.

- Muthukumarswamy, R., Revathi, G., and Vadivelu, M. (2000). *Acetobacter diazotrophicus* - Prospects and Potentials: An overview, In: *Recent Advances in Biofertilizer Technology* (editors, Yadav et. al.), pp. 126–153.

- Muthukumarswamy, R., Revathi, G., Seshadri, S., and Laxshminarasimhan, C. (2002). Gluconacetobacter Diazotrophicus, A Promising Diazotrophic Endophytes in Tropics, Curr. Science, **83**, pp. 137–145.

- Myer, L., Okech, Bernard, A., Mwobobia, Isaac, K., Kamau, A., Muiruri, S., Mutiso, N., Nyambura, J., and Mwatele, C. (2008). Use of Integrated Malaria Management Reduces Malaria in Kenya, In Myer, Landon, PLOS ONE, **3**(12), p. 4050.

- Nayak, B. (2001). Uptake of Potash by Different Plants with the Use of Potash Mobilizing Bacteria (Frateuria aurantia), Thesis, QUAT, Bhubaneswar.

- Nayar, P. K., Mishra, A. K., and Patnik, S. (1982). Silica in Rice and Flooded Rice Soils, II Uptake of Silica in Relation to Growth of Rice Varieties of Different Durations Grown in an Inceptisol, Oryza, **19**, pp. 88–92.

- Neeru Narula (Editor) (2000). Azotobacter in Sustainable Agriculture, New Delhi. ISBN – 81-239-0661-7.

- Nianikova, G. G., Kuprina, E. E., Pestova, O. V., and Vodolazhskaia, S. V. (2002). Immobilizing of Bacillus Muciloginosus a Producer of Exopolysaccharides, on Chitin, *Prikladnaia biokhimiia J. Mikrobiologiya*, **38**, pp. 300–304.

- Nierazwicki-Bauer, S. A. (1990). Azolla-Anabaena symbiosis: Use in Agriculture, In Handbook of Symbiotic Cyanobacteria, Amar N. Rai (Editor), CRC Press, Boca Raton, Florida, pp. 119–136.

- Nora Schultz (2009). Photosynthetic Viruses Keep World's Oxygen Levels Up, New Scientist.

- Norkina, S. P., and Pumpyansakya, L. V. (1956). Certain Properties of Silicate Bacteria Dokl, *Crop Sci., Soc.*, Japan, **28**, pp. 35–40.

- Norris, D. O., Jensen, H. L. (1958). A Procelain Method for Storing of *Rhizobium.*, *J. Exp. Agric.*, **31**, pp. 255–258.

- Nunez, C., Moreno, S., Soberon-Chavez, G., and Espin, G. (1999). The Azotobacter Vinelandii Response Regulato AlgR is Essential for Cyst Formation, *Journal of Bacteriology*, **181**(1), pp. 141–148.

- Okon, Y. and Labandera-Gonzalez, C. (1994). Agronomic Applications of *Azospirillum*: An Evaluation of 20 Years Worldwide Field Inoculation, *Soil Biology and Biochemistry*, **26**, pp. 1591–1601.

- Ota, M., Kobayashi, H., and Kawaguchi, Y. (1957). Effect of Slag on Paddy Rice 2, Influence of Different Nitrogen and Slag Levels on Growth and Composition of Rice Plant, *Soil and Plant Food*, **3**, pp. 104–107.

- Padmaja, P. and Verghese, E. J. (1972). Effect of Ca, Mg, and Si on the Uptake of Plant Nutrients and Availability of Straw and Grain of Paddy, *Agri. Res. J. Kerala*, **10**, pp. 100–105.

- Page, W. J. and Sadoff, H. L. (1975). Relationship between Calcium and Uronic Acids in the Encystment of Azotobacter Vinelandiil, *Journal of Bacteriology*, **122**(1), pp. 145–151.

- Page, W. J., Tindale, A., Chandra, M., and Kwon, E. (2001). Alginate Formation in Azotobacter vinelandii UWD during Stationary Phase and the Turnover of Poly-ß- hydroxybutyrate, *Microbiolgy*, **147**(2), pp. 483–490.

- Park, M., Singvilay, D., Seok, Y., Chung, J., Ahn, K., and Sa, T. (2003). Effect of Phosphatesolubilizing Fungi on 'P' Uptake and Growth of Tobacco in Rock Phosphate Applied Soil, *Korean J. Soil Sci. Fertil.*, **36**, pp. 233–238.

- Parker, C. A. (1957). Evaluation of Nitrogen-fixing Symbiosis in Higher Plants, *Nature*, **179**, p. 593.

- Parker, L. T. and Socolofsky, M. S. (1968). Central Body of the Azotobacter Cyst., *Journal of Bacteriology*, **91**(1), pp. 297–303.

- Paula, M. A., Urguiaga, S., Siqueira, J. O., and Dobereiner, J. (1992). Synergistic Effect of Vesicular Arbuscular Mycorrhizal Fungi and Diazotrophic Bacteria on Nutrition and Growth of Sweet Potato, *Biol. Fertili. Soils*, **14**, pp. 61–66.

- Peng, G., Wang, H., Zhang, G., Hou, W., Lium Y., Wang, E. T., and Tan, Z. (2006). Azospirillum melinis sp. Nov., A Group of Diazotrophs Isolated from Tropical Molasses Grass, *Int. J. Syst. Evol. Microbiol.*, **56**, pp. 1263–1271.

- Peoples, M. B. and Herridge, D. F. (1995). Nitrogen Fixation by Legumes in Tropical and Subtropical Agriculture, *Advances in Agronomy*, **44**, pp. 155–221.

- Peoples, M. B., Herridge, D. F., and Ladha, J. K. (1995). Biological Nitrogen Fixation: An Efficient Source of Nitrogen for Sustainable Agricultural Production, *Plant Soil*, **174**, pp. 3–28.

- Phillips, J. M. and Hayman, D. S. (1970). Improved Procedure for Clearing Roots and Staining Parasites and Vascular Arbuscular Mycorrhizal Fungi for Rapid Assessment of Infection, *Trans. Br. Myco. Soc.*, **55**(1), pp. 158–161.

- Pillai, Kamalasananan, P., Premalatha, S., and Rajamony, S. (2008). Azolla-A Sustainable Feed Substitute for Livestock, Farming Matters Magazine.

- Pisciotta, J. M., Zou, Y., and Baskakov, I. V. (2010). Light-Dependent Electrogenic Activity of Cyanobacteria, In: Yang, Ching-Hong. PLOS ONE, **5**(5), p. 10821.

- Plenchette, C., Fortin, J. A., and Furlan, V. (1982). Growth Responses of Several Plant spp. to Mycorrhizae in a Soil of Moderate Soil P Fertility, *Plant and soil*, **70**, pp. 199–209.

- Plenchette, C., Perrin, R., and Duvert, P. (1989). The Concept of Soil Infectivity and A Method for its Determination as Applied to Endomycorrhizas, *Can. J. Bot.*, **67**, pp. 112–115.

- Pope, L. M. and Wyss, O. (1970). Outer Layers of the Azotobacter vinelandii Cyst, *Journal of Bacteriology*, **102**(1), pp. 234–239.

- Postgate, J. R. (1982). The Fundamentals of Nitrogen Fixation, Cambridge, United Kingdom, Cambridge University Press.

- Powell, C. L. (1980). Mycorrhizal Fungi Stimulate Uptake of Soluble and Insoluble Phosphate Fertilizer from a Phosphate Deficient Soil, New photologist, **80**, pp. 351–358.

- Powell, C. L. (1980). VA-mycorrhizal Inoculation of Field Crops, Proc. N.l. Agron. Soc., **12**, pp. 85–92.

- Preston, T. R. and Murgueitio, E. (2011). Sustainable Intensive Livestock Systems for the Humid Tropics, FAO.

- Pronk, J. T., Meulenberg, R., Hazeu, W., Bos, P., and Kuenen, J. G. (1990). Oxidation of Reduced Inorganic Sulphur Compounds by Acidophilic Thiobacilli, FEMS Microbiol. Rev., **75**, pp. 293–306.

- Rajaee, S., Alikhani, H. A., and Raiesi, F. (2007). Effect of Plant Growth Promoting Potentials of Azotobacter chroococcum Native Strains on Growth, Yield and Uptake of Nutrients in Wheat, Journal of Science and Technology of Agriculture and Natural Resources, **11**(41), p. 297.

- Rajan, S. (2002). Comparison of Phosphate Fertilizers for Pasture and their Effect on Soil Solution Phosphate, Comm. Soil. Sci. Plant Anal., **33**, pp. 2227–2245.

- Ramarethinam, S. and Chandra, K. (2005). Studies on the Effect of Potash Solubilizing/mobilizing Bacteria Frateria Aurantia on Brinjal Growth and Yield, *Pestol*, **11**, pp. 35–39.

- Rawlings, D. E. and Kusano, T. (1994). Molecular Genetics of Thiobacillus Ferrooxidans, *Microbiology and Molecualr Biology Reviews*, **58**(1), pp. 39–55.

- Reeves, F. B., Wagner, D., Moorman, T., and Kiel, J. (1979). The Role of Endomycorrhizae in Revegetation Practices in the Semi-arid West. I, A Comparison of Incidence of Mycorrhizae in Severely Distributed vs. Natural Environment, Am. J. Bot., **66**, pp. 6–13.

- Reinhold, B., Hurek, T., Fendrik, I., Pot, B., Gillis, M., Kersters, K., Thielemans, S., and De Ley, J. (1987). *Azospirillum halopraeferens* sp. Nov., A Nitrogen Fixing Organism Associated with Roots of Kallar Grass [*Leptochloa fusca* (L.) Kunth], *Int J Syst Bacteriol*, **37**, pp. 43–51.

- Richardo, S. O., Mariano, D. C., Dallacovg, R., Albuquexque, A. N., Laboto, A. K., Guedes, E. M., Oliveira, C. F., Conceicao, H. E., and Alves, G. A. (2013). Azospirillum: A New and Efficient Alternative to Biological Nitrogen Fixation in Grasses, *J. Food Agriculture and Environment*, **11**(1), pp. 1142–1146.

- Robert, F., Scagel, Rober, Bandoni, J., Glenn E. Rouse, Schofield, W. B., Janet, R., Stein, and Tylor, T. M. (1965). An Evolutionary Survey of the Plant Kingdom, Belmont, California, Wadsworth Publishing Co., Inc., p. 658.

- Robinson, E. and Robbins, R. C. (1968). Emissions Concentrations and Fate of Gaseous Atmospheric Pollutants, Menlo Park, California, Stanford Research Institute.

- Robinson, E. and Robbin, R. C. (1970). Gaseous Sulphur Pollutants from Urban and Natural Sources, *J. Air. Pollut. Control. Assoc.*, **20**(40), pp. 233–235.

- Robson, R. L., Eady, R. R., Richaardson, T. H., Miller, R. W., Hawkins, M., and Postgate, J. R. (1986). The Alternative Nitrogenase of Azotobacter Chroococcum is a Vanadium Enzyme, *Nature*, **322**(6077), pp. 388–390.

- Ruschel, A. P., Henis, Y., and Salatia, E. (1975). Nitrogen-15 Tracing of N-fixation with Soil Grown Sugarcane Seedling, *Soil Biol. Biochem.*, **7**, pp. 181–182.

- Russell, E. W. (1973). Soil Conditions and Plant Growth, 10th Edition Longman, London, p. 756.

- Ruttimann-Johnson, C., Rubio, L. M., Dea, D. R., and Ludden, P. W. (2003). VnfY is required for full activity of the Vanadium-containing dinitrogenase in Azotobacter vinelandii, *Journal of Bacteriology*, 185(7), pp. 2383–2386.

- Sadoff, H. L. (1975). Encystment and Germination in *Azotobacter* vinelandii, *Microbiolgical Reviews*, 39(4), pp. 516–539.

- Mehnaz, S., Weselowski, B., and Lazarovitis, G. (2007). Azospirillium zeae sp. nov., A Diazotrophic Bacterium Isolated from Rhizosphere Soil of Zeae Mays, *International Journal of Systematic and Evolutionary Microbiology*, 57, pp. 2805-2809.

- Saxena, M. M. (1989). Environmental Analysis of Water, Soil and Air, Agrobotanical publ. India, p. 786.

- Schofield, P. E., Gregg, P., and Syers, J. K. (1981). Biosuper as a Phosphate Fertilizer, A glass house evaluation, *Newzealand J. Exp. Agricult.*, 9, pp. 63–67.

- Schwencke, J., and Caru, M. (2001). Advances in Actinorrhizal Symbiosis: Host plant-Frankia Interactions, Biology, and Application in Arid Land Reclamation, A Review, *Arid Land Res. Mang.*, 15, pp. 285–327.

- Schwintzer, C. R., and Tjepkema, J. D. (editors) (1990). The Biology of Frankia and Actinorrhizal Plants, Academic Press, Inc., New York.

- Shank, Yu., Demin, O., and Bogachev, A. V. (2005). Respiratory Protection Nitrogenase Complex in Azotobacter Vinelandii, *Success Biological Chemistry*, 45, pp. 205–234.

- Sheng, X. F., and Huang, W. Y. (2002). Mechanism of Potassium Release from Feldspar Affected by the Strain NBT of Silicate Bacterium, *Acta. Pedol. Sin.*, 39, pp. 863–871.

- Sheng, X. F., and Huang, W. Y. (2002). Study on the Conditions of Potassium Release by Strain NBT of Silicate Bacteria Scientia, *Agriucltura Sinica*, 35(6), pp. 673–677.

- Sheng, X. F., Xia, J. J., and Chen, J. (2003). Mutagenesis of the Bacillus Edaphicus Strain NBT and its Effect on Growth of Chilli and Cotton, Agric. Sci., China.

- Sheng, X. F. (2005). Growth Promotion and Increased Potassium Uptake of Cotton and Rape by a Potassium Releasing Strain of Bacillus Edaphicus, *Soil Bio. Biochem.*, 37(1), pp. 1918–1922.

- Sheng, X. F., and Le, L.Y. (2006). Solubilization of Potassium Bearing Minerals by a Wild Type Strain of Bacillus Edaphicus and its Mutants and Increased Potassium Uptake by Wheat, *Canadian J. Microbiol.*, 52(1), pp. 66–72.

- Shi, D. J., and Hall, D. O. (1988). The Azolla-Anabaena Association: Historical perspective, symbiosis and energy metabolism, *The Botanical Review*, **54**, pp. 353–386.

- Shivprasad, S., and Page, W. J. (1989). Catechol Formation and Melanization by Na+-Dependent Azotobacter chroococcum: A Protective Mechanism for Aeroadaptation? *Applied and Environmental Microbiology*, **55**(7), pp. 1811–1817.

- Siefert, E., and Pfennig, N. (1979). Chemoautotrophic Growth of Rhodopseudomonas Species with Hydrogen and Chemotrophic Utilization of Methanol and Formate, *Arch. Microbiol.*, **122**, pp. 177–182.

- Silva, J. A. (1971). Possible Mechanisms of Crop Response to Silicate Applications, *Proc. Int. Symp. Soil Fert. Evaluation*, **1**, pp. 805–814.

- Simms, E. L., Taylor, D. L., Povich, J., Shefferson, R. P., Sachs, J. L., Urbina, M., and Tauszick, Y. (2006). An Empirical Test of Partner Choice Mechanisms in a Wild Legume-rhizobium Interaction, *Proc. Roy. Soc. of London*, B., **273**, pp. 77–81.

- Singh, C. S., and Subba Rao, N. S. (1979). Associative Effect of *Azospirillum* Brasilense with *Rhizobium* Japonicum on Nodulation and Yield of Soybean (Glycine max), *Plant and Soil*, **53**, pp. 387–392.

- Singleton, P.W., Swaify, S. A., and Bahool, B. B. (1992). Effect of Salinity on *Rhizobium* Growth Parameters, *Applied Enviromental Microbiology*, **44**, pp. 884–890.

- Sjodin, Erik (2012). The Azolla Cooking and Cultivation Project, ISBN: 978-91-98068-60-3.

- Socolofsky, M. D., and Wyss, O. (1962). Resistance of the Azotobacter Cyst, *Journal of Bacteriology*, **84**(10), pp. 119–124.

- Sommer, M. D., Fuzyakov, and Breuer, J. (2006). Silicon Pools and Fluxes in Soils and Landscapes – A Review, *Journal of plant nutrition and soil science*, **169**, pp. 310–329.

- Sorokin, D. Y., Lysenko, A. M., Mityushina, L. L., Tourova, T. P., Jones, B. E., Rainey, F. A., Robertson, L. A., and Kuenen, J. G. (2001). *Thioalkalimicrobium aerophilum* gen.nov., sp. Nov., and *Thioalkalimicrobium sibericum* sp. Nov., and *Thioalkalivibrio versutus* gen. nov., sp. Nov., *Thioalkalivibrio nitratis* sp. Nov., and *Thioalkalivibrio denitrificans* sp. Nov., novel obligately alkaliphilic and obligately chemolithoautotrophic sulfur-oxidizing bacteria from soda lakes, *Int. J. Syst. Evol. Microbiology*, **51**, pp. 565–580.

- Sperberg, J. I. (1958). The Incidence of Apatite Solubilizing Organisms in the Rhizosphere and Soil, *Australian J. Agril. Resou. Econ.*, **9**, p. 778.

- Sprent, J. I., and Sprent, P. (1990). Nitrogen Fixing Organism, Pure and applied aspects, London, United Kingdom: Chapman & Hall.

- Stamford, N. P., Silva, J. A., Freitas, A., Araujo, and Filho, J. T. (2002). Effect of Sulphur Inoculated with Acidithiobacillus in a Saline Soil Grown with Leucena and Mimosa Tree Legume, *Biores Technology*, **81**, pp. 53–59.

- Stephan, M. P., Oliveiza, M., Teixeira, K. R. S., Martinez-Drets, G., and Dobereiner, J. (1991). Physiology and Dinitrogen Fixation of *Acetobacter* Diazotrophicus, *FEMS microbiology letter*, 77(1), pp. 67–72.

- Stetter, K. O., Fiala, G., Huber, G., Huber, H., and Segerer, A. (1990). Hyperthermophilic Microorganisms, *FEMS Microbiology*, *Rev.*, **75**, pp. 117–124.

- Steve Nadis (2003). The Cells that Rule the Seas, Scientific American.

- Stewart, I., and Falconer, I. R. (2008). Cyanobacteria and Cyanobacterial Toxins, In: Oceans and Human Health: Risks and Remedies from the Seas; Waslh, P. J., Smith, S. L., and Fleming, L. E. (Editors), Academic Press, pp. 271–296.

- Strohl, W. R. (2009). Genus I. Beggiatoa, In: Bergey's Manual of Systematic Bacteriology [Staley, J. T., Bryant, M. P., Pfennig, N., and Holt, J. G. (Editors)], vol. **3**, Williams & Wilkins, Baltimore, Md.

- Styriakova, I., Styriak, I., Galko, I., Hradil, D., and Bezdicka, P. (2003). The Release of Iornbearing Minerals and Dissolution of Feldspar by Heterotrophic Bacteria of Bacillus Species, Ceramic. Silicaty, 47(1), pp. 20–26.

- Styriakova, I., Bhatti, T. M., Bigham, J. M., Styriak, I., Vourniess, A., and Tuoviness, O. H. (2004). Weathering of Phlogopite by Bacillus Cereus and Acidiothiobacillus Ferroxidans, *Canadian J. Microbiology*, **50**, pp. 213–219.

- St. John, T.V. (1990). The Role of Mycorrhizae in Plant Ecology, *Canadian Journal of Botany*, **61**, pp. 1005–1014.

- Subba-Rao, N. S. (1980). Crop Responses to Microbial Inoculation, In: Recent advances in nitrogen fixation, London, United Kingdom: Edward Arnold, pp. 406–420.

- Subba-Rao, N. S., Tilak, K. V. B. R., and Singh, C. S. (1985). Synergistic Effect of Vesicular-arbuscular Mycorrhizae and *Azospirillum brasilense* on the Growth of Barley in Pots, *Soil Biology and Biochemistry*, **17**, pp. 119–121.

- Sugumaran, P., and Janarthanam, B. (2007). Solubilization of Potassium Containing Minerals by Bacteria and their Effect on Plant Growth, *World J. Agric. Sci.*, **3**(3), pp. 350–355.

- Sulvia, D. M., and Burks, J. N. (1988). Selection of a Vesicular-arbuscualr Mycorrhizal Fungus for Practical Inoculation of Uniola Paniculata, mycologia, **80**(4), pp. 565–568.

- Sundara, B., Natarayan, V., and Hari, K. (2001). Influence of Phosphorus Solubilizing Bacteria on Soil Available P. status and Sugarcane Development on a Tropical Vertisol, *Proc. Soc. Sugarcane Technology*, **24**, pp. 47–51.

- Supanjani, H. S., Han., Jung, S. J., and Lee, K. D. (2006). Rock Phosphate Potassium and Rock Solubilizing Bacteria as Alternative Sustainable Fertilizers, *Agro. Sustain. Develop.*, **26**, pp. 233–240.

- Suylen, G. M. H., Stefess, G. C., and Kuenen, J. G. (1986). Chemolithotrophic Potential of a Hyphomicrobium Species, Capable of Growth on Methylated Sulphur Compounds, *Archives of Microbiology*, **146**, pp. 192–198.

- Sylvia, D. M., and Jarstfer, A. J. (1990). Sheared Root Inoculums of Vesicular Arbuscular Mycorrhizal Fungi, *Applied Environmental Microbiology*, **58,** pp. 229–232.

- Tabatabai, M. A. (1986). Sulphur in Agriculature, *American Society of Agronomy*, Madison, WI, p. 428.

- Tandon, H. L. S. (1987). Phosphorus Research and Agriculture Production in India, Fertilizer development and consultation organization, New Delhi, p. 160.

- Takahashi, E., and Miyake, Y. (1982). The Effect of Silicon on the Growth of Cucumber Plant, Proc. 9th Inter Pl. Nutr. Collog., Warwick University, UK, p. 669.

- Tarrand, J. J., Krieg, N. R., and Dobereiner, J. (1978). A Taxonomic Study of the Spirillum Lipoferum Group, with Descriptions of a New Genus Azospirillium gen. nov and two Species, Azospirillium lipoferum (Beijerinck) comb. Nov. and Azospirillium brasilense sp. Nov. *Candian J. Microbiol.*, **24**, pp. 967–980.

- Taun, D. T., and Thuyet, T. Q. (1979). The Use of *Azolla* in Rice Production in Vietnam, In: International Rice Research Institute, Nitrogen and Rice, Los Banos, Philippines, pp. 395–405.

- Tejera, N., Llulch, C., Martinz-Toledo, M. V., and Gonzalez-Lopez, J. (2005). Isolation and Characterization of Azotobacter and Azospirillum Strains from the Sugarcane Rhizosphere, *Plant and Soil*, **270** (1–2), pp. 223–232.

- Tchan, Y. T., and New, P. B. (1986). Genus I Azotobacter, Beijerinck (1901), In: Bergy's Manual of Systematic Bacteriology [Krieg, N. R., and Holt, J. (Editors)] Williams and Wilkins Cu. Baltimore, **1**, pp. 220–229.

- Tepper, E. Z., Shilnikova, V. K., and Pereverzev, G. I. (1979). Workshop on Microbiology, M. P, p. 216.

- Theodorou, M. E., and Plaxton, W. C. (1993). Metabolic Adaptations of Plant Respiration to Nutritional Phosphate Deprivation, *Plant Physiology*, **101**, pp. 339–344.

- Thiagalingam, K., Silva, J. A., and Fox. R. L. (1977). Effect of Calcium Silicate on Yield and Nutrient Uptake in Plant Growth on a Humic Ferriginous Latosol, In: *Proc. Conf.on chemistry and fertility of tropical soils*, Kuallalumpur, Malaysia, Malaysian Society of Soil Science, pp. 149–155.

- Thuler, Daniela. S., Floh, E. L., Handro, W., and Heloiza, R., Barbosa (2003). Beijerinckia derxii releases plant growth regulators and amino acids in synthetic media independent of nitrogenase activity, *Journal of Applied Microbiology*, **95**(4), pp. 799–806.

- Tilak, K. V. B. R. (1993). Bacterial Fertilizers, ICAR, New Delhi.

- Tisdale, S. L., Nelson, W. L., and Beaton, J. D. (1985). Soil Fertility and Fertilizers, MacMillan Publ., New York, p.754.

- Tisdale, S. L., Nelson, W. C., Beaton, J. D., and Havlin, J. L. (1993). Soil Fertility and Fertilizers, 5th Edition, McMillon Publishing Co., New York.

- Tomich, T., Kilby, P., and Johnson, B. (1995). Transforming Agrarian Economics: Opportunities Seized, Opportunities Missed, Itheca, Cornell University Press.

- Truper H. G., and Fischer, U. (1982). Anaerobic Oxidation of Sulfur Compounds as Electron Donors for Bacterial Photosynthesis, Phil. Trans. R. Soc., London, **298**, pp. 529–542

- Tulasne, L. R., and Tulasne, C. (1841). Observations Sur le genre Elaphomyces, Et description de quelques especes nouvelles, *Ann. Sci. Nat. Ser.*, Z., **16**, pp. 5–29

- Ullman, W. J., Kirchman, D. L., and Welch, W. A. (1996). Laboratory Evidence by Microbial Mediated Silicate Mineral Dissolution in Nature, *Chem. Geol.*, **132**, pp. 11–17.

- Vajda, V., and McLoughlin, S. (2005). A New Maastrichtian-paleocene *Azolla* Species from Bolivia with a Comparison of the Global Record of Coeval *Azolla* Microfossils, Alcheringa: *Australasian Journal of Palaeontology*, **29**(2), pp. 305–329.

- Van Hove, C., and Diara, H. F. (1987). *Azolla* Production in African Agriculture, Progress and Problems, *International Rice Commission Newsletter*, **36**, pp. 1–4.

- Vassilev, N; Medina, A; Azcon, R and M. Vassilev (2006) Microbial solubilization of rock phosphate as media containing agro industrial wastes and effect of the resulting products on plant growth and phosphorus uptake. **Plant Soil** 287: 77-84.

- Vaughan, Terry (2011). Mammalogy (http://books.google.com).

- Vela, G. R., and Rosenthal, R. S. (1972). Effect of Peptone on *Azotobacter* Morphology, *Journal of Bacteriology*, **111**(1), pp. 260–266.

- Verma, V. K. (2011). Studies on On-farm Production Techniques of Arbuscular Mycorrhizal Fungi, M. Sc. thesis submitted to MPKV, Rahuri, p. 60.

- Vessey, K. J. (2003). Plant Growth Promoting Rhizobacteria as biofertilizers, *Pl. Soil.*, **255**, pp. 571–586.

- Vinitikova, H. (1964). A Contribution to Study on Efficiency of Silicate Bacteria, *Rostl. Vyr.*, **37**, pp. 1219–1228.

- Vittadini, C. (1842). Monographie Lycoperdineum. Milan [Also published, 1843, in Mem. R. Acad. Sci. torino, **2**(V), p. 216].

- Wagner, G. M. (1997). Azolla: A Review of its Biology and Utilization, *The Botanical review*, **63**, pp. 1–26.

- Wallace, W. H., and Gates, J. E. (1986). Identification of Eubacteria Isolated from Leaf Cavities of Four Species of the N-fixing *Azolla* fern as *Arthrobacter* Conn and Dimmick, *Applied Environmental Microbiology*, **52**, pp. 425–429.

- Wani, P. V. (1980). Studies on Phosphate Solubilizing Microorganism, A Review, *Journal of Maharashtra Agric. Univ.*, **5**, pp. 144–147.

- Watanabe, I. (1982). Azolla-Anabaena Symbiosis, its Physiology and Use in Tropical Agriculture, pp. 169–185. In: Dommergues, Y. R. and Diem, H. G. (Editors), *Microbiology of tropical soils and plant productivity*, Martinus Nighoff/W. junk, The Hague.

- Watanabe, I., and Van Hove, C. (1996). Phylogenetic, Molecular, and Breeding Aspects of Azolla-Anabaena symbiosis, In: J. M. Camus, M. Gibby and R. J. Johns (Editors), *Pteridology in Perspective*, Royal Botanic Gardens, Kew., pp. 611–619.

- Weier, K. L. (1980). Nitrogen Fixation Associated with Grasses, Tropical Grassland, **14**(3), pp. 194–201.

- Welser Anzeiger (1921). The Rapid Compost Methods by Robert Raabe, Berkeley en. Wikipedia.org/wiki/compost.

- Whitelaw, M. A., Harden, T. Y., and Bender, G. L. (1997). Plant Growth Promotion of Wheat Inoculated with *Penicillium radicum* sp. Nov., *Australian. J. Soil Res.*, **38**, pp. 291–300.

- Will, M. E., and Sylvia, D. M. (1990). Interaction of Rhizosphere Bacteria, Fertilizer, and Vesicular Arbuscular mycorrhizal Fungi with Sea Oats, *Applied and Environmental Microbiology*, **56**, pp. 2073–2079.

- Wong, T. Y., and Maier, R. J. (1985). H2-Dependent Mixotrophic Growth of N2-Fixing Azotobacter vinelandii, *Journal of Bacteriology*, **163**(2), pp. 528–533.

- Wu, S. C., Cao, Z. H., Li, Z. G., Cheung, K. C., and Wong, M. H. (2005). Effect of Biofertilizer Containing N-fixer, P and K Solubilizers, and AM-fungi on Maize Growth, *Geoderma*, **125**, pp. 155–166.

- Wyss, O., Neumann, M. G., Socolofsky, M. D. (1961). Development and Germination of the *Azotobacter* cyst., *Journal of Biophysical and Biochemical Cytology*, **10**(10), pp. 555–565.

- Xie, C. H., and Yokota, A. (2005). *Azospirillum oryzae* sp. Nov., A Nitrogen-fixing Bacterium Isolated from the Roots of the Rice Plant *Oryza sativa.*, *Int J. Syst Evol Microbiology*, **55**, pp. 1435–1438.

- Xue, Q. H., Sheng. J. W., and Tang, L. (2000). Effect of K Bacteria on Nutrients Activation in Lou Soil, *Acta Agriculture Boreali-Occidentalis Sinica*, **9**(3), pp. 67–71.

- Yahalom, E., Okon, Y., and Dovrat, A. (1987). *Azospirillum* Effects on Susceptibility to *Rhizobium* Nodulation and on Nitrogen Fixation of Several Forage Legumes, *Canadian Journal of Microbiology*, **35**, pp. 510–514.

- Yamagata, U., and Itano, A. (1923). Physiological Study of *Azotobacter* chroococcum, beijerinckii, and vinelandii types, *Journal of Bacteriology*, **8**(6), pp. 521–531.

- Zachmann, J. E., and Molina, J. A. E. (1993). Presence of Culturable Bacteria in Cocoons of the Earthworm Eisenia fetida, *Applied and Environmental Microbiology*, **59**(6), pp. 1904–1910.

- Zhang, C. J., Tu, G. Q., and Cheng, C. J. (2004). Study on Potassium Dissolving Ability of Silicate Bacteria, *Shaguan College Journal*, **26**, pp. 1209–1216.

- Zimmerman, W. J. (1985). Biomass and Pigment Production in Three Isolates of *Azolla* II, Response to Light and Temperature Stress, *Ann. Bot.*, **56**, pp. 701–709.

- A Brief History of Solid Waste Management (http://www.stormcon.com).

- Preventing contaminants in home compost piles (2012) (http://www/networx.com).

- Edmonton composting facility (http://www.edmonton.ca).

- Frankia infection process (2011) (http://web.uconn.edu).

- Frankia nitrogen fixation (2011) (http://web.uconn.edu).

- Frankia taxonomy (http://web.uconn.edu).

- *Franki alni*: A symbiotic nitrogen-fixing actinobacterium.

- (http://www.genoscope.cns.fr)

- Robert Raabe, The Rapid Compost Method, Berkeley (http://vric.lucdavis.edu).

A

Acetobacter 8, 28
Acetobacter diazotrophiucs 29, 82
Aceylene Reduction Assay 68
Algal biofertilizers 135
Algal flakes 134
Anabaena azolla 15
Arbuscular mycorrhizal fungi 136
Associative Nitrogen fixing
Biofertilizers 30
Azolla 3, 8, 12, 90, 136
Azolla circinata 12
Azolla filiculoides 12
Azolla japonica 12
Azolla Mexicana 12
Azolla microphylla 12, 16, 90
Azolla nilotica 12
Azolla pinnata 12
Azospirillum 8, 30, 79
Azospirillum brasilense 33
Azospirillum Canadense 33
Azospirillum lipoferum 33
Azospirillum sp. 30
Azotobacter 8, 16, 18, 75
Azotobacter sp. 19

B

Bacillus mucilaginosus 48
Bacillus polymyxa 42
Bacterial fertilizers 6
Bacterial inoculants 6
Beggiatoa 53

Beijerinckia 20
Beijerinckia indica 8
BGA 3, 6, 8, 23, 26, 28, 87
Biofertilizers 3, 4, 6
Bioinoculant 6
Biological Nitrogn fixation 2, 3, 4

C

Chlorobium limicola 8
Chromatium minus 8
Clostridium pasteurianum 8
Cyanobacteria 8, 24, 25, 26

D

Decomposing cultures 6, 59, 119
Derxia gummose 8
Desulphovibrio sp. 8
Drip Irrigation 147
Duckweed fern 12

E

EM solutions 61

F

Fairy moss 12
Feldspars 57
Fertilizer N 2
Fertilizer use efficiency 4
Fertilizers 1
Foliar spray 147
Frankia 34, 35, 36, 97
Frankia alni 36, 37, 38

G

Gigaspora calospora 105
Glomus epigaeum 105
Glomus mosseae 105
Growth promoting substances 73

L

Liquid biofertilizer 128, 147

K

Kjeldanl methold 8
Klebsiella pneumonia 8

M

Methanobacterium sp. 8
Micas 57
Micro S-109 54
Microbial cultures 6
Mosquito fern 12
Mycorrhizal dependency 106

N

Nitragin 4
Nitrogen fixing biofertilizers 6
Nitrogenase 19
Non-symbiotic nitrogen fixing
Biofertilizers 8, 16

O

Organic decomposer 3

P

Phosphate solubilizing
Biofertilizers 7, 39, 42
Phosphate solubilizing microbes 3,
    42
Phosphate solubilizing
Microorganism 99
Potassium solubilizing
Biofertilizers 7, 46, 117
Production of compost 138

R

Red azolla 13
Rhizobium 3, 8, 9, 63
Rhizobium japonicum 9
Rhizobium leguminosarum 8
Rhizobium lupine 9
Rhizobium melioti 9
Rhizobium phaseoli 8
Rhizobium trifoli 9
Rhodospirillum capsulatus 8
Rhodospirillum rubrum 8

S

Seed pelleting 147
Seed treatment 145, 146
Seedling treatment 145, 147
Sett treatment 146
Silica solubilizing bacillus sp. 58
Silicate solubilizing biofertilizers 7,
    54, 113
Soil broadcasting 147
Soil treatment 146, 147
Starter culture 130
Sulphur oxidizer 3
Sulphur oxidizing bacteria 109
Sulphur oxidizing biofertilizers 7,
    50
Symbion-S 54
Symbiotic nitrogen fixing 8

T

Thiobacillus bioftertilizer 54
Thiobacillus thiooxidans 54

V

Vesicular arbuscular
Mycorrhiza 3, 43, 103

Z

Zinc & sulphur solubiliser 3

Author index

A

Abd-Elmonem and Amberger 40
Ahmad & Ahmad 20
Ahmad et.al. 20
Aleksandrov et. al. 47, 48
Allaby 23
Allen 44
Al-Sherif 3
Aragno 51
Armstrong 39
Arnold 12, 14
Avakyan et. al. 47
Arveby and Huss-Danell 38

B

Badr 47, 50
Bagyaraj 45
Baillie et. al. 17
Baker & O' Keefe 98
Barber 39
Barbosa et. al. 21, 22
Bashan 30
Bashan and Holguin 30, 31
Bashan and Levanony 30, 31
Bashan et. al. 31
Becking 21, 22
Beijerinck 16
Bellenger et. al 19
Ben Dekhil et. al. 30
Berg 38
Berkum & Bohlool 17

Berthelin 48
Bertsch et. al. 47
Bethlenfalvay et.al. 45
Bocchi and Malgioglio 12
Boddey and Dobereiner 29, 30
Bowen et.al. 44
Boyd & Boyd 16
Brinkhuis et al 14
Brown et.al. 30
Brune 51, 52
Bumb 1
Burgmann 19
Burris 2

C

Callaham et. al. 35
Carrapico 13
Carrapico et. al. 12
Carrapico and Tavares 12, 13
Cavalcante and Doberenier 28, 29, 30
Chandra et. al. 49
Chein 40
Chein et.al. 40, 41, 42
Chen et.al. 20
Chinnasami & Chandrasekaran 57
Christophe et. al. 49
Ciobanu 56
Clarson 49
Cohen et. al. 14, 24
Costa 12
Curatti et.al. 19

D

Datta & Shinde 57
Datta et. al. 48
De los Rios 23
De zwart et. al. 51
Dela Cruz et. al. 45
Denison 11
Derx 21
Dicker & Smith 17
Diem and Dommergues 36
Dixon & wheeler 2
Dobereiner 21, 28
Dong et. al. 57
Drozdowicz and Ferrira Santos 31
Durrant et.al. 19

E

Eckert et.al. 30
Ehrlich 54
Elawad et. al. 56
Emitiazia et.al. 20
Epstein 57
Eriksson 53

F

FAO 1, 2
Forni et. al. 13
Francis & Reid 103
Friedrich 51
Friedrich & Mitrenga 51, 52
Friedrich et. al. 47, 48
Friend 53
Fuchs et. al 51
Fuentez-Ramirez et.al. 30
Funa et.al. 17

G

Galindo et al. 20
Gallo et. al. 56
Gama-Castro et.al. 17
Gandora et al. 16
George & Garrity 18

Gerdemann 103
Gerdemann & Nicolson 103
Gibson et. al. 68
Gillis and de ley 28
Gillis et.al. 30
Giraud 10
Goldstein 46
Granat et. al. 53
Grey 77
Gromov 47
Grover 41

H

Halsall and Gibson 31
Han and Lee 49
Han et. al. 50
Hans Guntel Schleget et. al. 20
Harley and Smith 44
Harrison 51
Heath & Tiffin 11
Heckman 59
Heinen 47
Hill et.al. 20
Hiraishi & Umeda 52
Holguin and Bashan 31, 32
Hopper 1
Howard and Rees 19
Hu et. al. 19, 47
Huber and Stetter 51
Hung & Sylvia 106
Hung et. al. 103
Hussner 12

I

IFA 1
Imhoff et. al. 52
Isopi et al. 31

J

Jaga and Patel 1
James et. al. 29
Jarstfer & Sylvia 105
Jasper et.al. 45

Jensen 19
Johnstone 16
Jones 17

K

Kaneshiro et. al. 68
Kannaiyan 2, 13
Kass et. al. 19
Kellogg et. al. 53
Kelly 51
Kelly & wood 51
Kelly and Smith 51
Kelly et.al. 51, 53
Keyser and li 11
Khammas and Keiser 31
Khammas et al. 30
Khudsen 48
Kiers 11
Koide and Mooney 45
Kondratieva 52
Kovalcic et.al. 45
Krajick 38
Krieg and Holt 48
Kumar et al. 16

L

Lalonde and Quispel 38
Layne & Johnson 17
Lea-smith et.al. 25
Lee 23
Lee et. al. 30
Lewis 17
Li 47
Li et.al. 20
Libbert and Risch 30
Lin & Sadoff 17
Lin et.al. 18, 48
Loperfido and sadoff 18
Lowry et. al. 68
Lumpkin and Plucknett 12

M

Macy et.al. 51

Magalhaes et.al. 30
Maier and Moshiri 19
Marco et. al. 11
Martyniuk & Martyniuk 16
Michiels et. al. 31
Mikhailouskaya and Tchernysh 47, 49
Miller and Eady 19
Mishustin & Shilinikora 77
Moira et. al. 48
Money et.al. 20
Moore 14
Moorman and Reeves 45
Moreno et. al. 16
Muralikannan 57
Muralikannan & Anthoni Raj 57
Murry et. al. 98
Muthukumarswamy et al. 29, 30
Myer et. al . 15

N

Nayak 48
Nayar et. al. 56
Neeru 20
Nianikoval et. al. 48
Nierzwicki-Bauer 13
Nora Schultz 23
Norkina & pumpyanskaya 47
Nunez et. al. 17

O

Okon and Labander- Gonzales 30
Ota et. al. 57

P

Padmaja & Verghese 56, 57
Page & Sadoff 17
Page et.al. 20
Park et. al. 49
Parker and Socolofsky 17
Paula et al. 29, 30
Peng et.al. 30
Peoples et.al 2

Philips & Hayman 107, 108
Pillai et al 15
Pisciotta et.al. 25
Plenchette et. al. 106, 107
Pope & Wyss 17
Postgate 2
Powell 45
Preston & Murgueitio 15
Pronk et al. 51

R

Rajan 41, 42
Rajee et.al. 20
Ramarethinam and Chandra 49
Rawlings and Kusano 51
Reeves et.al. 44
Reinhold et.al. 30
Ricardo et.al. 32
Robert et. al. 14
Robinson & Robbins 53
Robson et.al. 19
Ruschel et al. 28
Russel 56
Ruttimann-Jonson et.al. 19

S

Sadoff 17
Saxena 114
Schofield et.al. 40
Schwencke and Caru 38
Schwintzer and Tjepkema 35, 38
Shank et.al. 19
Sheng 49
Sheng & Huang 46, 47
Sheng and Le 48, 50
Sheng et. al. 48, 49
Shi & Hall 13
Shivprasad and page 19
Siefert and Pfenning 52
Silva 56
Simm et. al. 11
Singh & Subba Rao 31
Singleton et. al. 63

Sjodin Erik 15
Socolofsky & wyss 17
Sommer et. al. 55
Sorokin et.al. 51
Sperberg 47
Sprent & sprent 3
St.John 45
Stamford et.al. 41
Stefano Bocchi & Antonino
Malgio-glio 26
Stephan et al. 28
Stetter et. al. 51, 53
Steve Nadis 24
Stewart and Falconer 26
Strasburger 12
Strohl and Genus 51
Styriakova et. al. 48, 49
Subba-Rao 2, 31
Sugumaran and Janarthanam 47, 50
Sundara 40
Sundara et.al. 40
Supanjani et. al. 50
Suylen et. al. 52
Sylvia Jarstfer 106

T

Tabatabai 41
Takahashi & Miyake 56
Tandon 39
Tarrand et.al. 30
Taun and Thuyet 12
Tchan & New 77
Tejera et. al 16, 80
Tepper et. al. 19
Theodorou and plaxton 39
Thiaglingam et. al. 55
Thuler et. al. 22
Tilak 8
Tisdale et.al. 39, 40, 56
Tomich et al. 1
Truper and Fischer 51
Tulasne and Tulasne 43

U

Ullman et. al. 48

V

Vajda and McLoughlin 12
Van Hove & Diara 13
Vassilev et.al. 50
Vaughan Terry 23
Vela & Rosenthal 17
Vessey 40, 49
Vinitikova 57
Vittadini 43

W

Wagner 12, 13, 14
Wallace & Gates 13
Wani 42
Watanabe 12
Watanabe and Van Hove 13

Weier 21
Welser Anzeiger 59
Whitelaw 40
Will and Sylvia 31
Wong & Maier 18
Wu et. al. 49
Wyss et. al. 18

X

Xie & Yokota 30
Xue et. al. 48

Y

Yahalom et.al. 31
Yamagata & Itano 16

Z

Zachmann & Molina 17
Zhang et. al 49
Zimmerman 14